Countering Counterfeit Trade

Thorsten Staake • Elgar Fleisch

Countering Counterfeit Trade

Illicit Market Insights, Best-Practice Strategies,
and Management Toolbox

 Springer

Thorsten Staake
ETH Zurich, Department of Management,
    Technology, and Economics
Sonneggstrasse 63 (SOW F 11)
8092 Zurich
Switzerland
tstaake@ethz.ch

Elgar Fleisch
ETH Zurich, Department of Management,
    Technology, and Economics
University of St. Gallen,
    Institute of Technology Management
Dufourstrasse 40a (ZIG)
9000 St. Gallen
Switzerland
elgar.fleisch@unisg.ch

ISBN 978-3-642-09562-7   e-ISBN 978-3-540-76947-7

Cover design: KünkelLopka Werbeagentur, Heidelberg, Germany

Printed on acid-free paper

9 8 7 6 5 4 3 2 1

springer.com

Foreword

Trade in counterfeit goods has developed into a substantial threat to many industries. The problem is no longer confined to prestigious, easy-to-manufacture products which consumers all too often knowingly purchase as cheap imitations. Today counterfeiting affects pharmaceuticals, mechanical spare parts, fast-moving consumer goods and electronic components as well as fashion accessories, clothing, cigarettes, and digital media. Even within the different product categories counterfeit supply is extremely diverse. While some goods pose a severe risk to the health and safety of consumers, others have a decent level of quality and satisfy the needs of most of their users; manufacturing sites exist where people work under almost inhuman conditions whereas other facilities resemble modern, highly automated plants; and counterfeit producers may act like small-time criminals, but may also manage their businesses like multinational companies. On the demand-side consumers sometimes invest considerable effort in searching for low-cost imitations or may become actively engaged in a company's anti-counterfeiting program.

The implications for brand owners are as manifold as the market itself. Effects include losses of revenue due to substitution and constraints to product pricing, erosion of brand value when corresponding goods appear to become less exclusive or of lower quality, liability claims and product recalls if substandard imitations end up in products or on the shelves of licit companies, and increased competition resulting from learning effects among illicit actors who may eventually turn into licit manufacturers. However, some brand owners can also benefit from trademark infringements. Especially in emerging markets, imitations are often like barriers to market entry for local competitors and can furthermore familiarize a large number of people with the product.

In fact, the phenomenon and its implications are highly complex. Counterfeiting is a ruthless crime as well as a smart knowledge-transfer strategy with benefits for at least a subset of consumers – and it is everything in between. It is just not possible to explain it with oversimplified assumptions, and concentrating solely on its criminal elements and the danger to society may lead to overlooking some important characteristics that are helpful when defining anti-counterfeiting strategies.

In our book we do not promise to tell you everything you need to know to stop intellectual property infringements. That would be like offering a way to consistently outperform the stock market. However, we will provide you with a thorough analysis of the supply- and demand-side of the illicit market, outline state-of-the-art brand- and product-protection strategies of successful companies, and introduce a set of tools to determine the financial implications for individual companies. Furthermore, we will highlight some major problems with existing product security technologies and discuss what benefits and hurdles brand owners can expect when using Radio Frequency Identification (RFID) technologies to protect their goods. Unlike many other publications in this field, our book is not meant as a wake-up call. It is directed at those who have already recognized the extent of the problem as well as the implications for consumer safety and company value. Some findings – for example from the market share estimates and impact analyses – may not always suit the IP lobbyist, but we believe that taking an impartial approach is the best way to support management decisions and to convince government organizations to take action. In fact, we see our work's benefits in the unbiased analysis of the counterfeit market, the strong focus on assisting practitioners to deal with the challenges, and the high standard of research that backs up the findings presented. We hope to support those who are currently implementing or improving their anti-counterfeiting strategies, aim to offer some new and insightful perspective for long-serving veterans in this field, and try to stimulate an ongoing learning process that is necessary to successfully respond to counterfeit actors.

Table of Contents

Methodology

This book combines the findings from four years of academic research with hands-on experience of anti-counterfeiting projects across different industries. For the development of effective brand- and product-protection measures, both sources proved to be equally important. From our perspective, the practical work was crucial to highlight the deficiencies of existing brand- and product-protection strategies, to describe working solutions, and to identify best practice approaches. The lessons learned from many industry experts constitute an indispensable asset in the fight against illicit actors. In fact, the projects with leading companies from the luxury consumer goods, fast-moving consumer products, aviation, and the pharmaceutical industry helped us to truly understand the problem and to ensure the solutions' relevancy to practice.

The academic research, on the other hand, enabled us to challenge the established approaches and to develop strategies that go beyond the state-of-the-art procedures. Though counterfeit trade is a relatively new field for academia, many valuable contributions can be found in adjacent areas – e.g. on unlawful behavior, consumer choice and substitution effects with respect to genuine products, the perceptions of brands, innovation strategies, competitor analyses, production settings, etc. – that one can build upon. A systematic analysis, for example of the business cases of illicit actors and the barriers of exit and entry for engaging in counterfeit trade, can help to find ways to effectively reduce counterfeit supply – rather than just reacting to counterfeit occurrences. A "gut feeling" on the illicit market may be misleading, and a solid methodology is crucial for obtaining substantiated insights. The same is true for market share and impact analyses that all too often only vaguely reflect the actual situation, as well as for surveys on consumer behavior that need to be carefully designed in order to ensure a reasonable level of validity. Throughout this book, we refer to related academic publications and draw on established research methods for deriving and validating our findings. This approach is essential to shed light on the clandestine phenomenon with its criminal actors, hidden sales channels, and multifaceted consumer behavior on the one side and the interests of the brand owners on the other. We encourage the reader to follow this critical, impartial way of thinking as it very likely helps to develop the most effective measures for protecting the consumers and safeguarding the companies' assets.

Acknowledgements

This book is based on numerous applied research projects conducted at the Swiss Federal Institute of Technology (ETH) Zurich, the University of St. Gallen, and the Massachusetts Institute of Technology. It reflects the state-of-the-art of the academic work on counterfeit trade, and the findings are deeply rooted in theory. What we regard as its key feature, however, is the combination of theory with managerial practice. This would have not been possible without our industry partners and clients. We especially like to thank a number of individuals and organizations who supported us along the way.

First, we want to express our gratitude to the members of the Special Interest Group (SIG) Anti-Counterfeiting. In this working group, which we set up in 2004, our partners from Bundesdruckerei, Gillette (now Procter & Gamble), Infineon Technologies, Montblanc, SAP, and Swarovski provided valuable insights into this highly industry-specific problem. Their engagement and willingness to challenge the current understanding of counterfeit trade greatly helped to advance the knowledge of the illicit market and thus to develop effective anti-counterfeiting strategies. Over time, the activities around the SIG have developed into a vivid field of research, covering many aspects from counterfeit supply and demand over market share and impact analyses to organizational and technical countermeasures.

The anti-counterfeiting activities were taken to the next level in 2006 when the European Commission decided to co-fund the project SToP with a research grant of over EUR 2.8 million. We want to express our gratitude to our project officer, Dr. Peter M. Friess, who supported our work and also encouraged us to investigate those aspects which are not yet on the agenda of most organizations. Furthermore, we want to thank the members of the consortium, Airbus, Bundesdruckerei, Novartis, Oria Computers, Richemont, SAP, and Spacecode, for the fruitful and trustful collaboration.

Together with Schering (now Bayer Schering Pharma), we initiated a cross-industry benchmarking of anti-counterfeiting measures. We especially want to thank Dr. Stephan Schwarze and Marina Bloch of Bayer Schering Pharma, who contributed greatly to the success of this project. In an unparalleled empirical analysis, more than 50 leading companies across different industries participated in the empirical study and provided valuable insights into the state-of-the-art of their brand- and product-protection efforts.

The work has also been put forward by the Auto-ID Labs within the Flagship Project Anti-Counterfeiting and Supply Chain Security. We would like to thank Prof. Sanjay E. Sarma and Prof. John R. Williams from the Massachusetts Institute of Technology and Prof. Peter H. Cole from the University of Adelaide for the vivid exchange of ideas and their feedback throughout the numerous projects, and Chris Adcock and Nicholas Fergusson of EPCglobal for their advice, financial support, and the many doors they have opened for us.

The work also stood to benefit from the exchange with several colleagues. We wish to thank Prof. Dr. Friedemann Mattern for his outstanding support; Dr. Christian Flörkemeier for his numerous valuable insights into Radio Frequency Identification (RFID) and our discussions on product counterfeiting; Dr. Frédéric Thiesse and Dr. Florian Michahelles for the very good teamwork on the numerous projects; Dr. Christian Tellkamp and Prof. Dr. Stephan Billinger for our discussions on research methodology in our fields; Matthias Lampe for the development of the Auto-ID Product Check Demonstrator; Dr. Zoltan Nochta for his contributions to the SIG, Mikko Lehtonen and Tobias Ippisch for their excellent contributions to many applied research and consulting projects, Glauco Degan, Steven Hirschbühl, and Prof. Dr. Thomas Friedli for their support of the benchmarking project; Joanna Niederer for the editing and proofreading; and last but not least Elisabeth Vetsch-Keller for her outstanding organizational support.

Thorsten Staake and Elgar Fleisch
Zurich, February 2008

Organization of the Book

The book is organized in five parts. In Part A we explain what counterfeit trade is and how it has developed, and we discuss in great detail the mechanisms behind illicit supply and demand. Based on this background information, Part B outlines the development of successful brand- and product-protection measures. We describe established anti-counterfeiting strategies from various industries, and this is followed by a discussion on how to adjust, extend, and implement these measures. In Part C we explain what cardinal effects – positive and negative – counterfeit trade has at a microeconomic level. Furthermore, we introduce an easy-to-use set of tools to structure an impact analysis. Part D is dedicated to product-protection technologies which are an integral part of many anti-counterfeiting efforts. We show how companies can apply security technologies, how counterfeit producers respond to the use of such technologies, and what brand owners can expect from Radio Frequency Identification (RFID) technology. The book closes with Part E, where we provide a summary of the managerial steps to be taken when defining an anti-counterfeiting strategy. We conclude with an outlook on the likely further development of the illicit market. The content of the individual chapters is briefly outlined below.

Part A Knowing the Enemy – The Mechanisms of Counterfeit Trade

Chapter 1 An Introduction to Counterfeit Markets

Chapter 1 outlines the development of counterfeit trade from a phenomenon that primarily affected luxury consumer goods to an "illicit industry" that produces a wide range of products at various levels of quality. The drivers and enablers behind this trend are summarized. Following the basic description of the situation, we highlight the deficits of the current understanding of counterfeit activities and the shortcomings of established countermeasures. In order to facilitate the further discussion on illicit activities, a set of definitions of counterfeiting, piracy, gray markets, factory overruns, and trafficking is introduced and the differences between deceptive and non-deceptive cases are pointed out.

Chapter 2 Understanding Counterfeit Supply

While companies often have a detailed knowledge of the strengths and weaknesses of their licit competitors, counterfeit supply appears to be a black box for many stakeholders. Only very few brand owners are aware of the strategies of counterfeit producers, while most reduce their intentions to making quick profits, which is clearly oversimplified. Addressing this shortcoming, Chapter 2 outlines different business cases of illicit actors and shows the various production settings of counterfeit producers. In fact, an empirical investigation reveals several generic types of strategies that are commonly followed by illicit actors. Each of these strategies gives rise to distinctive production capabilities, shipment, and selling tactics and has characteristic vulnerabilities that brand owners can leverage.

Counterfeit actors appear to resemble licit enterprises as they at least implicitly perform investment-risk-return considerations and are likely to only engage in counterfeit activities if the business case is more attractive than alternative activities (for example counterfeiting products of less-protected brands). We discuss the cost drivers from a counterfeiter's perspective and thereby reveal what measures of brand owners and enforcement agencies have the greatest prospects of driving illicit actors out of the market. The supply-side investigation is complemented with an analysis of the illicit distribution channels and a review of the most important academic contributions in this domain.

Chapter 3 Counterfeit Demand and the Role of the Consumer

Consumers may buy counterfeit goods knowingly or in the belief that they are purchasing genuine products. They may even try to ensure that they only obtain original articles – or invest considerable effort to acquire less expensive fakes. In fact, understanding their multifaceted roles is essential for evaluating the implications of counterfeit trade on licit enterprises and for developing effective countermeasures. Chapter 3 provides insights into the consumers' awareness with respect to the existence of counterfeits in various product categories. The willingness to purchase counterfeit goods is analyzed, and reasons for and against intentional purchases are identified.

The survey-based findings enable licit manufacturers to assess counterfeit-related risks due to the lack of awareness on the side of the consumer and help to identify those who are likely to intentionally purchase illicit goods. Furthermore, the investigation of consumers' reasoning for and against intentional purchases helps to identify arguments to effectively influence public opinion on counterfeit purchases. Beyond presenting the results from the study, we also show how companies can obtain similar insights with respect to their own products and close with a summary of several insightful scholarly contributions on counterfeit demand.

Part B   Countermeasures – Best Practices and Strategy Development

Chapter 4 Established Anti-counterfeiting Approaches – Best Practices

Limiting trade in counterfeit goods is a common and non-conflicting goal of most brand owners. However, companies are individually struggling to develop and implement efficient countermeasures, often lacking the opportunity to learn from more experienced organizations or to compare the efficiency of their effort against other measures. Given the considerable interest in best practices, we describe the characteristics of successful monitoring, reaction, and prevention measures. The findings are based on a benchmarking study of more than 40 leading companies from various industries.

Research on managerial countermeasures has received some attention from academia. We therefore briefly summarize the most important findings and provide reference for those who wish to dig deeper into the details.

Chapter 5 Implementing Anti-counterfeiting Measures

The study presented in Chapter 4 not only provided insights into successful anti-counterfeiting measures but also revealed the limitations of existing approaches. This chapter shows how these shortcomings can be addressed. It provides concise guidelines for implementing effective monitoring, reaction, and prevention processes. We describe the different company-internal and external sources of information and explain how companies can combine these sources to obtain reliable, timely data on counterfeit occurrences, shipment strategies, sales channels, and the people and organizations producing and selling counterfeits. Furthermore, we discuss how companies can define reaction measures to respond to counterfeit occurrences as well as prevention strategies to secure the company's supply chain, eliminate production of counterfeit products, hamper their distribution, and discourage users or consumers from purchasing faked goods.

Part C Management Tools – Towards a Fact-based Managerial Approach

Chapter 6 Determining the Market Share of Counterfeit Articles

Data on the extent of counterfeit trade constitutes the baseline of any substantiated risk and impact analysis. However, no sound methodology to derive such estimates has been published yet. Chapter 6 highlights the most important problems with respect to existing market estimates. The discussion reveals that almost all frequently cited estimates severely overstate the market share of counterfeit goods. Concrete examples of erroneous analyses include studies on the share of counterfeit digital media, toys, pharmaceutical products, aviation spare parts, as well as macroeconomic calculations. We identify the pitfalls when assembling such market information. To obtain better data, a new computational framework is introduced. The framework resembles a sink-source model, makes use of multiple data sources and combines supply- and demand-side estimations to increase the accuracy of the results. It guides the data collection, provides error margins for the individual flows and the entire share of goods, and can be easily integrated in continuous monitoring activities. Exemplary macroeconomic calculations are presented, the maximum share of counterfeit goods among world merchandize trade is estimated, and the framework is applied to two brands.

Chapter 7 Implications for Affected Enterprises

Companies are hardly able to assess the impact of counterfeit trade on their business. Decisions regarding investing in countermeasures are likely to depend on "a gut feeling" rather than on a solid return on investment calculation – or may not be finalized due to a lack of financial data which could justify them. Addressing this issue, we provide a detailed analysis on the effects of imitation products on revenue, brand value, cost of quality, and future competition.

We start with an analysis of substitution effects among genuine and counterfeit articles. A simple but powerful model to determine the immediate loss of revenue due to faked products is developed. The model is backed up by a consumer survey on purchase decisions in markets with genuine and counterfeit goods. Sample calculations are provided for a luxury consumer product and a fast-moving consumer product. Following the revenue calculations, the implications of illicit imitation products on different functions of brands are discussed and a tool to quantify these effects is introduced. On a qualitative level, the effects on quality management and the risk of additional liability claims are assessed. Moreover, learning effects among illicit actors are investigated.

Counterfeit trade is not always bad for the manufacturer or brand owner. In fact, several positive effects may occur, mostly resulting from a higher perceived market share and enhanced accessibility in lower-price segments. These implications can be ascribed to the following categories: positive brand-related effects, network effects, and lock-in effects. We explain these effects and provide examples of how brand owners achieved a better market position by leveraging the existence of imitation products. Chapter 7 closes with a summary of the academic publications on the impact of counterfeiting.

Part D Product-protection Technologies

Chapter 8 Principles of Product Security Features

Holograms, flip colors, and micro printings are all prominent examples of established protection mechanisms. However, these static features constitute an ever-lower barrier for illicit actors, and many imitations today already resemble their genuine counterparts so closely that their inspection becomes a time-consuming process. Other more secure features such as chemical and biological markers are often not suitable for large-scale testing – but in a market where an increasing number of counterfeit goods intermingle with mass produced, genuine items, large samples or even complete checks are necessary. Against this background Chapter 8 provides an overview of the most important security features. Furthermore, an attack model is introduced to structure the requirements analysis. We show that threats not only result from weaknesses related to authentication (i.e., the duplication of features) but in particular from constraints during the inspection process, such as lack of resources for a careful investigation. We explain in greater detail the following attacks: cloning of features, obfuscation (confusion), tag omission, removal-reapplication and denial-of-service. Following this analysis the requirements for an RFID-based solution design are discussed.

Chapter 9 The Potential of RFID for Brand- and Product-protection

Severe drawbacks of the established anti-counterfeiting measures are the poor degree of standardization and automation for checking the authenticity of goods. Common security features mostly rely on a visual, object-by-object inspection which is impractical if goods arrive packaged or in bulk. Radio Frequency Identification helps to overcome this problem as it allows for bulk identifications without line-of-sight connection. Chapter 9 provides an introduction to RFID and the vision behind the "Internet of Things". The technical fundamentals of RFID transponders, readers and the underlying IT infrastructure are detailed. We discuss read ranges and read rates of RFID tags as well as the basic building blocks of the emerging EPC Network. Following the general introduction four RFID-based anti-counterfeiting approaches and three application scenarios are described in greater detail. We discuss the benefits and potential hurdles of each RFID implementation and outline the implications for monitoring, reaction, and prevention activities as well as the consequences for the different types of counterfeit producers.

Part E   Managerial Guidelines and Conclusions

In Chapters 10 and 11 we recapitulate the findings on counterfeit supply, consumer behavior and anti-counterfeiting strategies, and combine them into one overall guideline for practitioners. The book concludes with an outlook on the likely future development of the supply- and demand-side of the counterfeit market, government responses, and future organizational countermeasures.

Part A   Knowing the Enemy – the Mechanisms of Counterfeit Trade

1 An Introduction to Counterfeit Markets

Intangible assets, such as goodwill and intellectual property, constitute a significant share of many companies' equity. They are often the result of extensive investment in research and development, careful brand management, and a consistent pledge to high-quality, reliability, and exclusiveness. However, the growing momentum of emerging markets in Asia where these intangible assets are difficult to protect, a general trend in favor of dismantling border controls to ease the flow of international trade, and the increasing integration and interaction between organizations in disparate locations require new measures to protect these assets and safeguard companies from unfair competition.

Product counterfeiting in particular – the unauthorized manufacturing of articles which mimic certain characteristics of genuine goods and which may pass themselves off as registered products of licit companies – has developed into a severe threat to consumers and brand owners alike. The Organisation for Economic Co-operation and Development estimates that 5% to 7% of world merchandize trade is in goods infringing trademark, copyright and related rights.[1] Though this estimate is not supported by substantiated aggregated data and may be on the high side, it nevertheless expresses the magnitude of the problem. In an economy where many processes rely on extremely low failure rates, where single counterfeit cases can significantly influence public opinion on products and brands, and where organizations heavily rely on the protection of their intellectual property, even a fraction of 1% of counterfeit products can have serious implications on consumer safety, future competitiveness, and company profit.

Even more alarming than the sheer size of the counterfeit market is the increasing share of potentially dangerous and technically sophisticated imitation products. While the number of articles seized by European customs as well as the domestic market value of counterfeit goods confiscated at U.S. borders grew at an average annual rate of 11% between 2002 and 2006 (c.f. Info Box 1.1), the number of incidents that potentially led to physical injuries and caused expensive product recalls has skyrocketed. Alongside the traditional counterfeit articles such as designer clothing, branded sportswear, fashion accessories, tobacco products, and digital media, counterfeiting is having an increasing effect on a broader range of goods. Customs statistics show a considerable growth in trademark infringements among consumer products as well as among semi-finished and industrial goods. Today a considerable share of the cases includes counterfeit foodstuffs, pharmaceuticals, fast-moving consumer goods, electrical equipment, mechanical

---

[1] In fact, this estimate is highly questionable, as even the original source admits (OECD 1998). We provide an in-depth discussion on this issue in Chapter 6, where we show that the counterfeit market share among world merchandize trade is more in the order of 1% to 2%.

spare parts and electronic components (c.f. Info Box 1.2) – all in all articles from product categories where, due to the related risks to health and safety, European and North American customers would hardly purchase the counterfeit versions knowingly. In these so-called deceptive cases they often wrongly ascribe the poor quality of an imitation to the licit manufacturer.

While the number of counterfeit articles that are misattributed to the brand owner is of major concern especially in the western economies, the situation in emerging markets, in particular in China, is somewhat different. There many imitation products are manufactured and sold openly as cheap alternatives to the more expensive genuine products. Though most of these articles are not up to western quality standards, they often have at least some functional value and compete with their genuine counterparts in selected market segments. Both phenomena – misattributed imitations and competition with counterfeiters – have numerous severe implications for licit manufacturers and brand owners.

Info Box 1.1: Counterfeit seizures at customs

In Europe and the United States, customs inspects between 3% and 6% of all goods that flow into the country or common market. No other agency has access to a comparable amount of goods. Customs statistics are an important source of information on the development of the illicit market. The figure on the left shows the number of counterfeit articles seized by EU customs between 2000 and 2006, the figure on the right the domestic market value of goods confiscated by U.S. Border Controls between 2002 and 2006.

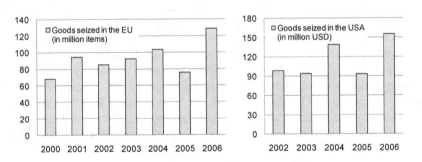

Both figures exhibit an average annual growth of 11%. However, as with many other statistics, this trend has to be interpreted with care. The development is to be seen in conjunction with the overall growth in world merchandize trade, customs officials' rising awareness of the problem, their enhanced skills in identifying intellectual property rights infringements, and the engagement of brand owners that is shown, for example, in an increasing number of applications for actions that are filed.

Info Box 1.2: Counterfeit incidents – breakdown by product type

On behalf of the European Commission, the Taxation and Customs Union publishes annual reports on the community of customs investigations into counterfeiting and piracy on the European border (EC 2007). Provided are the number of cases and the number of articles seized within different product categories, here shown for the year 2006.

Product type	No.of cases	% of total	No.of articles	% of total
Foodstuffs and beverages	54	0,1%	1.185.649	0,9%
Perfumes and cosmetics	1.093	2,9%	1.676.409	1,3%
Clothing and accessories	24.297	65,1%	14.361.867	11,2%
a) sportswear	3.254	8,7%	1.210.196	0,9%
b) other clothing	9.977	26,7%	4.315.338	3,4%
c) clothing accessories	11.066	29,6%	8.793.123	6,8%
Electrical equipment	1.342	3,6%	2.984.476	2,3%
Computer equipment (hardware)	543	1,5%	152.102	0,1%
Media (audio, games, software)	2.880	7,7%	15.080.161	11,7%
Watches and jewellery	3.969	10,6%	943.819	0,7%
Toys and games	678	1,8%	2.370.894	1,8%
Cigarettes	300	0,8%	73.920.446	57,5%
Medicines	497	1,3%	2.711.410	2,1%
Other	1.682	4,5%	13.287.274	10,3%
Total:	37.334	100,0%	128.631.295	100,0%

What is truly alarming is the amount of potentially (or better: very likely) dangerous imitation products. The number of counterfeit articles within the categories of foodstuffs and beverages, electrical equipment, and pharmaceutical products that are actually seized is in the order of millions per year – and many more pass the border without being inspected.

Another aspect that can be observed is that the number of registered cases seems to be only vaguely related to the number of seized articles in each category. Clothing and accessories account for more than 65% of the cases but only for 11.2% of the articles, while cigarettes make up 0.8% of the cases but more than 57% of all items seized. This shows that the lot sizes and thus also the import tactics vary considerably with respect to counterfeit trade in different industries. For a thorough interpretation, however, it is necessary to know how "one article" is defined. The allocation base may be obvious for handbags, less clear for cigarettes (one bar, one pack, or, as done in this report, one individual cigarette), and almost arbitrarily for foodstuffs (one consumption unit). Different approaches sometimes make comparisons between different countries difficult. Some stakeholders are interested in stressing the importance of the topic, while others try to play down the problem – and both have a sufficient degree of freedom to present the data accordingly.

Implications of counterfeit trade

Counterfeiting undermines the beneficial effects of intellectual property rights and the concept of brands as it affects the return on investment in research, development and company goodwill. Producers of reputable products are deterred from investing within a national economy as long as their intellectual property is at risk. The national tax income within the developed countries is reduced since counterfeit goods are largely manufactured by unregistered organizations. Social implications result from the above-mentioned costs as society pays for the distorted competition, eventually leading to less innovative products, higher taxes, unemployment, and a less secure environment as the earnings from counterfeiting are often used to finance other illegal activities (ICC 2005). However, for emerging markets, counterfeiting can also constitute a significant source of income and employment as well as an important element of an industrial-learning and knowledge-transfer strategy. As a consequence not all governments are determined to prosecute counterfeiters, which often renders legal measures in such markets ineffective.

For consumers the risks are considerably more imminent, mainly due to the possible health and safety hazards resulting from products of inferior quality. As the Commission of the European Communities states, "one of the most alarming dimensions of this phenomenon is the increased risk faced by EU citizens as a result of the growth in dangerous fake goods such as medicines, car parts and foodstuffs" (EC 2005a). Consequently, consumers often have a strong interest in buying genuine goods, especially when a registered trademark is seen as a sign of quality and thus helps to reduce search costs and purchasing risks. On the other hand, certain consumers buy counterfeit goods knowingly, especially when they regard the brand or trademark as an interpersonal sign of wealth and social status. A detailed understanding of such often ambivalent consumer behavior is essential when developing appropriate countermeasures as it determines whether they – not necessarily intentionally – support the licit brand owner or counterfeit producers.

For companies counterfeit trade can lead to: (1) a direct loss of revenue, since counterfeit products partially replace genuine articles; (2) a reduction in the companies' goodwill, because the presence of counterfeit products can diminish the exclusiveness of the affected brands and the perceived quality of a product; and to (3) a negative impact on the return on investment for marketing, research and development expenditures, which can result in a competitive disadvantage compared to those enterprises that benefit from free-ride effects. Moreover, counterfeit trade can (4) result in an increasing number of liability claims due to defective imitation products (especially in the case of health and safety hazards for consumers); and (5) facilitate the emergence of future competitors as counterfeiting can help illicit actors to gather know-how in production and may thus enable them to become licit enterprises in the future.

However, under specific circumstances counterfeit trade may also have some positive effects for the brand owners. A high counterfeit market share of counterfeit software products in emerging economies can, for example, constitute a barrier of entry for low-end competitors. At the same time imitations may familiarize a large user base with a product. Once intellectual property rights are more strictly enforced the market penetration of imitations is likely to translate into revenue for the brand owner.[2]

In most cases the associated risks and costs by far outweigh the benefits. The primary implications depend on the type of product, its applications, the risk associated with substandard articles, the cost of research and development, production, marketing, and the consumers' attitudes towards counterfeit articles within the product category under study. Figure 1.1 outlines the risk profile for companies as expressed by brand-protection specialists in the pharmaceutical, luxury goods, aviation and fast-moving consumer goods industries.[3]

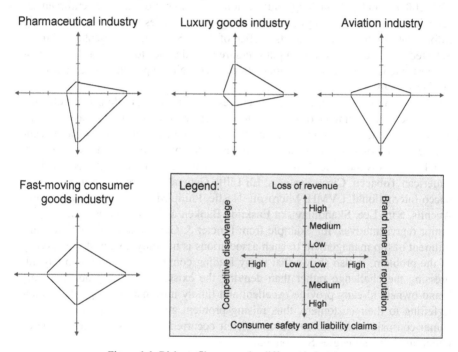

Figure 1.1: Risk profile as seen by different industries

[2] We provide a thorough discussion on the potentially positive effects and the ways to leverage them in Section 7.4.

[3] Overall, 22 brand-protection managers were asked to rate the perceived counterfeit-related risk on a scale of low, medium or high: "What is the perceived risk of counterfeit regarding the following four categories within your company?" (pharmaceutical industry N = 6; fast-moving consumer goods industry N = 6; aviation industry N = 3; luxury goods industry N = 7. The survey was conducted between 10/2005 and 10/2006).

Government and industry response – The relevance of the problem

Counterfeit trade has attracted considerable attention among trade associations, governments, and enterprises. Organizations such as the Business Action to Stop Counterfeiting and Piracy under the umbrella of the International Chamber of Commerce, the International Anti Counterfeiting Coalition, and the Union des Fabricates have become powerful advocates for a more stringent enforcement of intellectual property rights in international trade. In turn, governments of the developed nations, particularly the member states of the European Union, the United States and Japan, have stepped up their efforts to protect consumers and industry against counterfeit goods. Measures range from extending national border control (EC 2005b) and establishing multinational collaboration among enforcement agencies (EC 2006) to applying sanctions against countries that do not sufficiently enforce intellectual property rights (Office of the U.S. Trade Representative 2001). Selected regulatory bodies even put pressure on industries to use or at least evaluate cost-intensive security techniques to protect their supply chains against illicit products (FDA 2004).

The willingness of many companies to restrict counterfeit trade is well reflected by the increasing number of memberships in corresponding industry associations as well as by the commitment among senior management. Numerous senior executives chair or co-chair anti-counterfeiting associations, among them the Chairmen or CEOs of Vivendi, General Electric, Sony Corporation, Astra Zeneca, British American Tobacco, Cisco Systems, Eli Lilly, General Motors, Henkel, Japan Tobacco International, LVMH, Microsoft, Nestlé, Philip Morris International, Sanofi Aventis, Sara Lee, Skandinaviska Enskilda Banken and Unilever as well as other senior representatives, for example from Procter & Gamble and Pfizer. The commitment of top management to such associations is not only a sign of the severity of the problem, it also shows that many leading companies are now publicly addressing the challenge rather than denying the existence of the problem. Some brand owners already provide excellent and timely information on product counterfeiting to their customers, thus raising problem awareness and reducing consumer confusion when there are counterfeit occurrences. We will discuss related communication strategies Section 5.3.

1.1 The changing nature of counterfeit trade

Within the last decade the counterfeit market has changed dramatically. The development has affected not only the production capabilities of the illicit actors but also their logistics, sales and distribution activities. It is worth outlining these changes and identifying their underlying drivers in order to better predict the market's future development and to determine which factors licit actors should leverage to combat counterfeit trade.

Production

Counterfeiters benefited greatly from increasingly easy access to modern production facilities, from a larger number of skilled workers and from the growing demand within their domestic markets. The changing business conditions, however, did not lead to a homogeneous development as one might expect in licit markets but to a larger number of different production settings. The most important trends can be summarized as follows:

From easy-to-manufacture goods to a wide range of simple to sophisticated products. More than a decade ago clothing and fashion accessories dominated counterfeit supply. Having been limited in their access to capital and modern machinery, illicit actors almost exclusively used to harvest "low-hanging fruits", i.e. they tried to leverage brand-name-related free-rider effects while avoiding products that were complicated to manufacture. In fact, counterfeiters still often attach registered trademarks and logos either to generic products in order to sell them more or less openly as fakes or use imitations without any functional value in cases where the buyer cannot judge the quality of the products prior to purchase. Following this strategy, the marginal investments in production facilities and the limited impact of occasional seizures of equipment enable counterfeiters to operate highly profitably. This holds even if they have to sell obvious fakes far below the price of the original product.

While "logo counterfeiting" continues to exist, the targeted product categories today also include highly sophisticated products. When counterfeiting such goods, their producers are able to realize brand-name-related earnings and at the same time benefit from the original manufacturers' investment in research and development. However, sophisticated products often not only require extensive re-engineering capabilities but also call for expensive production machinery and a stable supply of semi-finished goods. Illicit actors therefore have to pay considerable attention to protecting their equipment, for example by creating a complex network of component manufacturers, by hiding illicit production behind licit activities, and by concentrating on those product categories where imitations are at least somewhat tolerated by local authorities. This commitment is often rewarded with considerable earnings from sales in their domestic markets.

From poor product quality to a wide range of quality levels. Following the increasing availability of production machinery, a growing number of counterfeit products that are marketed today have a decent quality. This does not mean that illicit imitations have become safer in general. As a matter of course counterfeit producers do not invest in rigid quality management. Especially in product categories where it is difficult to judge the value of an article prior to purchase, the functional quality is still mostly poor. Counterfeit drugs, for example, continue to pose a considerable threat to consumers and so do counterfeit food and alcoholic beverages. However, in emerging markets brand owners and licit manufacturers should especially focus on those products that already fulfill the basic needs of their buyers and compete with their genuine counterparts.

From a simple organizational structure in production to many different production settings. Industrial production requires an adequate organizational structure that reflects the complexity of the operation. In fact, a trend towards a professional, enterprise-like counterfeit organization can be observed, for example, with respect to the division of labor, (re-)engineering capabilities, and the established networks of suppliers, distributors and financial partners. However, positive business cases seem to exist for both low and high-end counterfeits and for large and small counterfeit organizations. In Section 2.1 we provide empirical evidence that several well-defined types of production settings exist, and that product complexity, product quality and the risk associated with counterfeit production are important factors to distinguish between the different types.

From low quantities to mass production. Global brands have become a key success factor not only in the luxury goods industry but also in industries with mass-produced goods. Well-known trademarks significantly reduce the search costs of customers and enable brand owners to realize considerable price premiums. Marketing expenses for building up the required level of brand awareness and the desired associations as well as the quality management that is necessary to consistently meet the expectations of customers amount to a considerable proportion of the overall product costs. Consequently well-known branded consumer goods developed into an interesting target for counterfeit producers when the required production, printing and packaging machinery became available. The trend has two detrimental effects in particular. Firstly mass-produced counterfeit goods pose a severe challenge to supply chain security measures, and secondly they are often sold as deceptive counterfeits, i.e. to customers who believe they are purchasing original products. The occurrence of mass-produced counterfeits is a major argument for frequent and cost-efficient product inspections.

From non-deceptive to deceptive counterfeiting. As indicated in the previous paragraph, the number of deceptive counterfeit cases has grown dramatically within the last decade.[4] In fact, deceptive counterfeiting is very attractive for illicit actors as they can fetch sales prices close to the prices of corresponding original products, while obvious fakes can only be sold in selected product categories and at significant discounts. The development towards deceptive counterfeiting is again facilitated by better access to advanced printing and packaging equipment.

---

[4] Deceptive counterfeiting is very difficult to observe. The statement is derived from an analysis of warranty cases in different industries. However, the estimated number of unreported cases is high.

Logistics

Illicit actors have to disguise the origin of their goods and minimize the risk of product seizures. As a consequence the shipment of illicit articles is often substantially more expensive than the distribution of genuine products. Alongside the advances in counterfeit production two major trends characterize the development of the corresponding logistics activities. They are firstly the growing importance of professional bootlegging networks for larger consignments, and secondly the use of postal services for small shipments.

Bootlegging networks. Counterfeits frequently apply tactics known from established bootlegging and drug-trafficking organizations. Measures to bypass border controls frequently include transshipment, i.e. the routing of shipments through countries that, in the past, neither have conducted effective inspections nor have been a significant source of counterfeit production and thus are not on the radar of customs officials in the country of destination. Other measures include admixing licit and illicit goods and hiding illicit goods in other shipments or vessels – very likely together with the whole spectrum from bribery to blackmail.

Establishing the network of actors while shielding different entities from each other (to ensure that one actor cannot endanger the adjacent parties) is expensive and time consuming. Though the associated costs are obviously justified given the extensive margins of most counterfeit products, shipment remains a weak point in the counterfeiters' value chain.

Small shipments. Especially in countries where intellectual property rights are strictly enforced and where street markets are not widespread, distribution to the end-customer constitutes a bottleneck for counterfeit supply. In that case postal services seem to become a popular distribution channel. Due to the sheer amount of mail illicit shipments are extremely difficult to identify at customs and thus seizure rates are low. Moreover, postal services do not require illicit actors to utilize their people networks and to share profits with intermediate stake holders. For brand owners the only promising approach to restrict this form of trade seems to be by limiting the demand, for example by informing potential consumers about the risks associated with ordering from clandestine sources.

Sales and marketing

When selling illicit goods, counterfeit actors have to strike the delicate balance between making it easy for their customers or victims to find and purchase the products (i.e. to reduce the search costs) and at the same time hiding the products they have manufactured from enforcement agencies and brand owners.[5] In fact,

---

[5] This statement holds for both deceptive and non-deceptive cases.

the difficulty to distribute illicit goods in many markets is still an important limiting factor of counterfeit trade, and the development of illicit sales activities have not kept pace with advances in production. The development of the most important sales channels is outlined below.

Street markets. The way counterfeit actors sell their products depends very much on the particular country. Street markets are dominant in the emerging economies, with their development reflecting the trends in counterfeit production (for example wider range of products etc.). In those western countries where street sales play no significant role, no significant changes have been observed with respect to most product categories. Here the high search costs of potential customers (due to the limited access such markets) significantly restrict the demand, as we will see in Section 3.1.[6]

The growing importance of the Internet for counterfeit sales. The Internet has developed into an important sales channel for counterfeit goods in Europe and North America. Especially pharmaceuticals and luxury goods are frequently advertised in unsolicited bulk emails (i.e. spam mails). What percentage of recipients respond to these emails has not yet been systematically analyzed, but many brand owners pay considerable attention to such sources (see Info Box 5.1 on page 89 for a description of tools to monitor counterfeit activities on the Internet).

Private imports. No significant change has been observed with respect to private imports by tourists who, while on holiday, purchase non-deceptive counterfeits or "remarkably cheap branded goods" and bring them home for personal use or as presents for friends and relatives. This flow of goods is sometimes described as ant-traffic, illustrating that even small lot sizes can add up to large quantities.

Licit supply chains. With a growing share of deceptive counterfeit articles that, by mere visual inspection, are very difficult to distinguish from genuine products, the distribution channels of licit companies are becoming an attractive target for illicit actors. Counterfeit consumer products are preferentially sold to small distributors at a discounted price that is often claimed to result from savings due to parallel imports or overproduction. Once in the licit supply chain the buyers rarely question the authenticity of the products.

Counterfeit parts in genuine products. Counterfeit parts occasionally end up as components in genuine products, leading to expensive recalls and highly visible counterfeit cases. If this happens, counterfeit parts have almost always been

---

[6] Counterfeit cigarettes are an exception. As cigarettes resemble "standard" products with only a few brands dominating the market, they meet with the demand of a large number of potential consumers who, due to considerable savings, are willing to accept the search costs or simply buy when having the opportunity. That explains the high market share despite the restrictions of the sales channel.

sourced from an unknown supplier, or have already been inbuilt in a component by a trustworthy but imprudent partner company. While the implications for manufacturers and brand owners can be huge, the overall importance of this sales channel from the counterfeiters' perspective is very likely only marginal.

Info Box 1.3: Counterfeit incidents – some examples

Electronic equipment. After its introduction in October 2001, Apple's iPod has become one of the most popular consumer electronic products ever. As of January 2008, more than 140 million units have been sold. Following this tremendous success, the first counterfeit cases became public in April 2006. These iPods look-alikes very closely resembled their genuine counterparts; they carried the Apple logo and were labeled with valid serial numbers. Both iPod shuffles and iPod nanos have been spotted. The differences to their genuine counterparts, as can be seen from the outside, are non-standard headphone jacks located on the lower-right and missing dock connectors. The counterfeit shuffle lacked a repeat setting and their battery light on the back. The packing was also quite convincing, but had the words "Digital Music Player" on the top, which the original does not. The devices did not meet the original's standards with respect to sound quality and battery life – but they were at least working and so difficult to distinguish from the original products that Apple warned its resellers of the existence of these fakes (McLean 2006). In fact, many owners are likely to believe that they have purchased a genuine product and will blame the brand owner for the poor performance of their device. This example strikingly shows that illicit actors are capable of producing even sophisticated goods, and that some of their products may even fulfill the basic needs of their customers.

Fast moving consumer products. In June 2007, the Colgate–Palmolive Company warned that toothpaste falsely packaged as "Colgate" had been found in several discount stores in New York, New Jersey, Pennsylvania, and Maryland. The product did not contain the expensive active ingredient fluoride, but it was contaminated with the harmful substance Diethylene Glycol. Several people in the eastern U.S. reported experiencing headaches and pain after using the product. Similar incidents occurred in Spain. The packages of the phony products have several misspellings including: "isclinically" "SOUTH AFRLCA" "South African Dental Assoxiation" (FDA 2007). From a brand-protection expert's view, the case is interesting for two reasons. First, it shows that well-known brands in the fast moving consumer goods industry are an attractive target even when the selling prices per volume and weight are low. Second, for counterfeiters, it seems to be sufficient to produce something that looks somehow similar to the original. Even goods where the brand serves as a sign of quality (rather than as a sign of social status) may carry obvious hints of their phony nature and still make it into otherwise licit stores.

Pharmaceuticals. When avian flu was a major topic in 2005, countries around the world were stockpiling the antiviral drug Tamiflu, patented and manufactured by the Swiss pharmaceutical company Roche, as a precaution against a possible pandemic. The concerns among U.S. citizens created a considerable additional demand for the medication, which even caused some shortages in supply. Already in November of 2005, customs seized the first counterfeit articles in a post office in South San Francisco. The pills did not contain the promised active ingredient, were shipped in small quantities, and were sold over the Internet (Walsh 2005). How many pills made it to the consumers is unclear as only a fraction of such small shipments are inspected by customs. The case made it very clear that illicit actors can very quickly leverage shortages in licit supply.

Counterfeit semiconductors. Shortly after the Tamiflu case, the Californian based semiconductor company QP reported another incident of product counterfeiting. The company, which specializes in military and other high reliability applications, found that a number of LM710 high-speed monolithic voltage comparators were fraudulent. Only due to extensive application-specific testing, were no final products with defective components shipped (O'Boyle 2006). Counterfeit or falsely labeled electronic devices are in fact a major problem; the frequent occurrence of such incidents shows that imitation products are not only sold to the final market but also threaten the parts supply of licit manufacturers.

1.2   A global problem – frequently discussed, little understood

Trade in counterfeit goods is a market phenomenon that takes place under specific basic conditions and involves stakeholders with characteristic interests and capabilities. Understanding these conditions, interests, and capabilities is essential for a sound evaluation of its implications and for the development of strategies to efficiently deal with counterfeit occurrences. However, though counterfeit trade is of considerable importance for numerous companies, the existing body of knowledge seems by no means to reflect the complexity of the illicit market. Existing anti-counterfeiting measures, including organizational and technological approaches, have not confined the recent growth of the counterfeit market. Open issues are itemized below:

The size of the counterfeit market. Estimating the extent of counterfeit trade appears to be a major challenge. Hardly any reliable statistics on this matter exist. Even the often cited and now widely accepted numbers provided by the OECD are highly questionable; no substantial aggregated data is available to support the high figures, as even the original source admits (OECD 1998 and 2006). Estimations of this kind may prove helpful to stress the importance of the topic, but are not sufficient as an input for detailed impact analysis or for decisions on the steps to be taken in specific geographic markets. It is virtually impossible for individual companies to leverage existing statistics as they neither apply suitable definitions of counterfeit trade nor provide the required selectivity for specific products. A survey among companies whose products are frequently seized by European customs revealed that less than 45% of the respondents possess plausible data regarding the share of counterfeits of their own products.

The financial impact on affected enterprises. Among the interviewed companies, only 7% of the respondents claimed to have reasonable estimates on the financial impact of counterfeit articles. Moreover, interviews concerning the impact on brand value revealed that the influences of counterfeit trade on brand loyalty, brand awareness, perception of quality and brand associations are only vaguely understood. Closely related to the lack of suitable impact analyses, only two out of ten manufacturers whose products are listed by European customs among the top 50 faked articles have defined indicators measuring the performance of their anti-counterfeiting activities. Without substantiated estimates concerning the market volume of counterfeit articles and suitable methodologies to translate these numbers into estimates of loss of revenue and goodwill, decisions on investments in countermeasures are likely to be based on "a gut feeling" rather than on a solid return on investment calculation – or may not be finalized due to a lack of justifying financial data. These deficiencies constrict the development and improvement of efficient monitoring, prevention, and reaction measures.

Risk assessments. In many industries the impact of counterfeit trade does not resemble a more or less constant financial loss but an exceptional event with potentially far-reaching consequences (for example in the aviation industry or for selected pharmaceutical products). In this case, scenario analyses and risk assessments are demanded by senior management to allocate the necessary resources for mitigating the risk. Conducting such analyses is a major challenge as neither the probability of occurrence nor the individual damage can be calculated in a straightforward way.

Strategies and production settings of illicit actors. The development of anti-counterfeiting measures requires some knowledge of the current situation of the illicit market (for example the share of counterfeit articles, current production locations and import routes, the quality of counterfeit products, etc.), as well as predictions about the future behavior of the illicit actors. Very few companies have been able to outline the potential strategies of their illicit competitors, and most respondents reduced the intentions of their opponents to "just realizing quick profits" – which is clearly oversimplified. Companies still seem to pay little attention to these industry-like mechanisms within the counterfeit market and often reduce the phenomenon to an opportunistic act rather than explaining it as a result of entrepreneurial considerations.

The role of the consumer. Counterfeiting can be seen as a disaggregation of brand and product. This may be detrimental for consumers who rely on brands as references to products with specific characteristics, but may also be desired by some of those for whom brands are of value by themselves (for example as a means to communicate wealth, social status, or membership in a certain group). Counterfeits of branded products that carry strong interpersonal value – Luis Vuitton, Rolex, or Montblanc, for example – are frequently purchased by consumers who are aware of the illicit nature of the article (non-deceptive counterfeiting). A differentiation between deceptive and non-deceptive counterfeiting is essential when determining the loss of revenue (the substitution effect for both cases isdifferent), the impact on brand name and quality perception (in the first case customers blame the licit manufacturer for the poor quality), and when developing anti-counterfeiting strategies (the buyer's willingness to help fight counterfeit trade depends on the reasons behind his or her intent to buy fakes). In this context important factors are consumer awareness and the willingness to purchase counterfeit goods. However, companies seem to lack the corresponding data and seem to have no access to appropriate methodologies to obtain such information.

The applicability of security technologies to fight illicit trade. Technological measures constitute an integral part of many anti-counterfeiting strategies. They serve as a means to authenticate genuine goods, help to distinguish them from counterfeits, and, for certain product categories, increase the production costs for illicit actors or confine the functionality of counterfeit articles. If properly deployed, technological measures strengthen the security of supply chains, hamper the production and distribution of counterfeit goods, and help to prevent the consumption of illicit articles. Holograms, flip colors, and micro printings are all prominent examples of established protection mechanisms. However, despite their high resistance against duplication, these features have not been able to stop the growth in counterfeit trade. In the above-mentioned survey only 41% of the respondents consider that established security features hold medium, high or very high prospects of successfully helping to avert counterfeit trade. The reasons for their systematic failure seem to be not fully understood and deserve further investigation.

1.3   Counterfeiting is not parallel trade is not overproduction – Why a clear problem definition is needed

Illicit trade denotes a wide variety of illegal or non-contractual activities. Trafficking in controlled substances, stolen and smuggled goods, trade of all kinds of products infringing intellectual property rights, and even parallel imports may fall into this category. Among these activities the enablers, the role of the actors, and the impact on affected enterprises are clearly different. However, many publications lump piracy and counterfeiting together with other forms of illicit trade, thus depriving themselves of the possibility to make use of the problems' unique characteristics (for example when analyzing the phenomenon or when developing countermeasures). In order to avoid potential inaccuracies, a working definition of counterfeit trade is developed in the remainder of this section.

Counterfeiting and piracy

The Agreement on Trade-Related Aspects of Intellectual Property Rights (the TRIPs Agreement) provides a widely-used definition of counterfeiting and piracy, which are both regarded as infringements of the legal rights of an owner of intellectual property. In greater detail,

- "counterfeit trademark goods" shall mean any goods, including packaging, bearing without authorization a trademark which is identical to the trademark validly registered in respect of such goods, or which cannot be distinguished in its essential aspects from such a trademark, and which thereby infringes the rights of the owner of the trademark in question under the law of the country of importation; whereas
- "pirated copyright goods" shall mean any goods which are copies made without the consent of the right holder or person duly authorized by the right holder in the country of production and which are made directly or indirectly from an article where the making of that copy would have constituted an infringement of a copyright or a related right under the law of the country of importation (WTO 1994).

Following this definition, the term "counterfeit goods" includes any goods bearing without authorization a trademark which cannot be distinguished in its essential aspects from the trademark registered for such goods, while "pirated goods" refers to infringements of copyright and related intellectual rights. However, breaches of trademark and copyright laws frequently overlap as companies often protect their products under either of the intellectual property rights. Therefore, in practice, the term "counterfeiting" encompasses any good which so closely imitates the appearance of protected products that it may mislead the buyer or a third person.

A working definition of counterfeiting

Consequently the following definition of counterfeiting is used in the remainder of this book: "Counterfeiting" denotes the unauthorized reproduction of goods, services, or documents in relation to which the state confers upon legal entities a statutory monopoly to prevent their exploitation by others. If referring to products, it can be defined as the manufacture of articles which are "intended to appear to be so similar to the original as to be passed off as genuine items" (c.f. WIPO 2004). This definition covers the illicit manufacturing process, but excludes other illicit activities such as bootlegging and trafficking in stolen products. The notion is in line with the French term "contrefaçon" and the German term "Produktpiraterie" (c.f. Clark 1997).

Deceptive and non-deceptive counterfeiting

Buyers may either acquire counterfeit articles in good faith – i.e. not being aware of the underlying intellectual property infringement – or may do so in full knowledge of the illicit nature of the product. In both cases the consumer behavior and sales tactics of the illicit actors are different, and both have specific implications for brand owners and manufacturers. For a differentiated discussion,

- *"deceptive counterfeiting"* shall refer to cases where a person or an organization purchases counterfeit goods in the belief they are buying genuine articles, and
- *"non-deceptive counterfeiting"* shall refer to cases where a person or an organization purchases counterfeit goods knowing of their counterfeit nature.

Parallel trade and overproduction

The terms "parallel trading" and "gray market activities" denote situations where goods are bought in one territory and distributed within another without the authorization of the right holder in the receiving market. Besides other effects, parallel trading limits the ability to realize regionally differentiated pricing strategies. It is facilitated by the principle of territoriality as intellectual property rights only apply to specific countries or economic areas, and the principle of exhaustion as right owners have limited power to control the distribution of products from the time they are legitimately put on the market. Enforcement agencies are mostly reluctant to take action against parallel traders (OECD 1998). In accordance with the previous definitions, parallel trade is not subsumed under the definition of counterfeit trade within this work.

The term "factory overrun" refers to unauthorized manufacturing by licit suppliers who can realize profits by producing extra quantities outside their license agreement. Though trademark owners often regard factory overruns as counterfeiting, and many industry associations include them in their counterfeit trade statistics, overruns constitute a breach of contract rather than a trademark infringement. Unless noted otherwise, we exclude overruns from the discussion as long as other product characteristics remain unchanged.

Info Box 1.4: Breakdown by type of rights of the number of articles seized

Customs accurately records the type of infringement that underlies each seizure act. The chart below shows the breakdown by type of infringement of the number of articles European customs confiscated in 2006. For merchandize trade, trademark infringements account for the largest share by far.

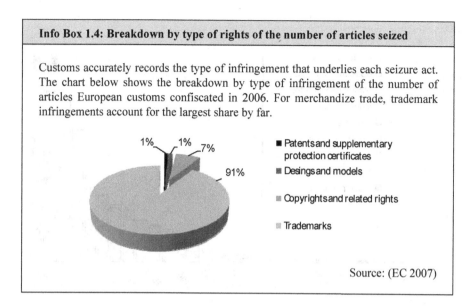

Source: (EC 2007)

1.4   Academic publications on counterfeit trade

Even though the first publications date back to the late 1970s, counterfeiting is still regarded as a rather young field of research. However, the public perception of counterfeit trade as a threat to both companies and consumers has risen dramatically in recent years, which is reflected by a large number of related publications in practitioner journals and the mass media (c.f. Figure 1.2). In parallel, the annual number of related scholarly contributions has risen as well. Counterfeiting research has not yet established itself as an autonomous research stream. Instead, it is distributed across different strands of management research, for example strategic management, marketing, logistics, and others. Against this background we have reviewed the existing body of academic literature so as to provide an integrated portrayal of the current level of knowledge in this field. We explicitly focused our analysis on journals in the area of management and economics. Accordingly, works in other disciplines such as law (for example contributions on the development of intellectual property rights in China) or engineering (for example contributions on the development of anti-counterfeiting technologies) are only mentioned if their individual contributions were of particular importance.

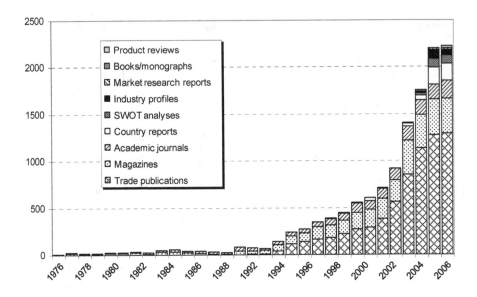

Figure 1.2: Counterfeiting-related publications in academic and practitioner journals 1976–2006 (retrieved from EBSCOhost Business Source Premier in December 2007)

The starting point for our analysis was an extensive search in electronic journal databases (Pro-Quest ABI/INFORM, EBSCOhost Business Source Premier) for the keywords "counterfeit", "counterfeiting", and "product piracy". In a second step we selected those contributions that concentrate on counterfeits in the narrow sense of the definition which we provided before. References from these studies were examined to identify further contributions from additional sources. Moreover, we decided to include selected reports from governmental authorities and industry associations, as these often provide the primary data that many other contributions build upon.

Some researchers have proposed the division of counterfeiting research into investigations of supply chain aspects on the one hand and demand-side aspects on the other (for example Bloch et al. 1993, Bush et al. 1989, and Tom et al. 1998). However, as will be shown later on, this classification does not do justice to the complexity of the subject, as many contributions are either too general or too focused on further issues to be assigned unambiguously to one of the two aspects. Therefore we propose to structure our review around the six categories depicted in Figure 1.3, considering the topic from different perspectives of the licit and the illicit supply chain as well as the interrelations between the two.

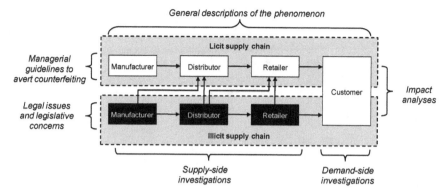

Figure 1.3: Overview of the counterfeit-related research streams

The major contributions from selected categories are discussed in greater detail later in this book:

- *Supply-side investigations* concern themselves with the production settings, tactics, and motives of illicit actors, and the ways in which their products enter the licit supply chain (see Section 2.4).
- *Demand-side investigations* focus on customer behavior and attitudes in the presence of counterfeit goods (see Section 3.3).
- *Managerial guidelines* to avert counterfeits comprise the tools and recommendations at organizational, strategic or technical levels for the management of affected companies (see Section 4.3).
- *Impact analyses* qualitatively investigate or quantify the consequences regarding turnover, brand value, liability claims and other key indicators for manufacturers of genuine goods and their supply chain partners (see Section 7.5).
- *Legal issues and legislative concerns* refer to different options for IP rights enforcement in the country of origin or in the respective market area to prevent – or at least to reduce – the availability of counterfeits goods (see last paragraph of Section 4.3).

2 Understanding Counterfeit Supply

Counterfeit producers are likely to base their operations upon – at least implicit – risk return calculations and on some sort of future planning or exit strategy. In fact, observations of the illicit market indicate that most actors follow concise strategies which are clearly reflected in a number of typical production settings and logistics activities. As with licit competitors, understanding the basic characteristics of counterfeit producers is crucial for shedding light on their strengths and weaknesses and for developing effective response and prevention strategies. In the following section we will provide an in-depth empirical analysis on counterfeit production for a wide range of product categories, discuss different business cases from an illicit actor's perspective and elaborate on distribution activities with respect to counterfeit supply.

2.1 Strategies and production settings of counterfeit producers

While the demand-side of the counterfeit market has received some attention in scholarly journals, very little is known about the market's supply-side.[7] In particular no research investigating the characteristics of counterfeit producers has been published, although a better understanding of this issue warrants attention for several reasons:

- First, the pervasiveness of counterfeit producers in many emerging economies with the associated income and learning effects is likely to influence the development of these nations (McDonald and Roberts 1994). Deeper insights into the supply-side of the counterfeit market can help to better understand their growth and progression, and may also improve strategies to protect intellectual property.
- Second, an empirical study regarding the basic characteristics of counterfeit producers can provide an anchor for further methodological investigations into this widely untouched field of business research. It may help to compare and contrast the existing body of knowledge with the new findings, for example probing theories and models on organizational learning, new venture strategies, customer value or brand management.
- Third, practitioners must select suitable strategies to protect their companies' revenue and intangible assets, sometimes even facing illicit markets similar in

---

[7] Insightful studies on counterfeit demand have been published for example by Grossman and Shapiro 1988a and 1988b, Bloch et al. 1993, Wee et al. 1995, Cordell et al. 1996, Chakraborty et al. 1997, and Gentry et al. 2006. We will discuss demand-side aspects in the following chapter.

size to their own and a local judicial system partly in favor of counterfeit producers (Ling 2005). Knowledge of the characteristics of competing illicit actors can prove helpful for decision makers in order to define and prioritize targeted countermeasures.

In the following empirical study we examine the characteristics of the counterfeit market. The study reveals the existence of five distinct strategic groups among counterfeit producers, (1) Disaggregators, (2) Imitators, (3) Fraudsters, (4) Desperados, and (5) Counterfeit Smugglers – each with specific characteristics with respect to production capabilities, re-engineering skills, properties of the targeted products, and potential degree of conflict with the law. While the work does not normatively address the performance or appropriateness of either group, it finds that the existence of each setting can be explained by strikingly simple, but not obvious, analytical considerations. The results allow for a more differentiated investigation of their learning and growth strategies, and also support practitioners to better position their companies with respect to the counterfeit market.

The considerations follow Porter's (1979) definition of strategic groups in respect to licit firms, and regard strategic groups as clusters of actors with similar strategies in terms of their key decision variables. Groups are separated from each other by mobility barriers, barriers to entry, and barriers to exit (Mascarenhas and Aaker 1989). These barriers can be skills and assets such as the ability to efficiently manufacture specific products, and – with respect to illicit behavior – also the ability to maintain illegal operations.

The context is the market of counterfeit goods (or more precisely the producers thereof) as observed between the years 2003 and 2006, where the narrow definition of counterfeit trademark goods applies as given in the Agreement on Trade-Related Aspects of Intellectual Property Rights. As outlined before, this definition comprises breaches of trademark law, but excludes piracy (which is defined as a violation of copyright and related rights), as well as factory overruns and parallel imports (which are considered a breach of contract rather than a breach of trademark law). It thus allows the focus to be set on the production activities of illicit actors.[8] The analysis is structured as follows:

- First, we describe the source of the underlying data and develop a set of group-defining variables including the measurement scale for each variable.
- Second, we describe the data sample and discuss the sample characteristics.
- Third, the actual cluster analysis is conducted, the results are presented, and the reliability and validity of the findings are discussed.
- Fourth, we conclude with the interpretation of the results.

---

[8] Products that infringe both copyrights and industrial property rights are as a matter of course included.

Data source and definition of the group-defining variables

The clandestine nature of the counterfeit market and the potential risks for counterfeit producers in case their identity is uncovered limits the direct accessibility to information from illicit actors. Though anecdotal evidence and testimonies from convicted actors constitute a potential source to validate selected findings retrospectively, counterfeiters who are able to hide their operations or who are tolerated by local enforcement agencies are likely to be underrepresented within this accessible group, thus introducing an indefinite sampling error. Moreover, the statements of convicted criminals are difficult to verify and therefore of limited value in the context of this study. Counterfeit products, however, provide valuable insights into the characteristics of illicit manufacturers. Expert analyses of such articles allow conclusions to be drawn on the re-engineering capabilities, the properties of and investment in corresponding production facilities, the functional quality of the products, and consequently on the likely strategic positioning of a counterfeiter's venture. For the present study, seized counterfeit articles serve as the primary data source of the empirical analysis where seizures were conducted by customs or other enforcement agencies, or resulted from test purchases by the right holders or licit manufacturers.

The choice of variables along which to group observations is a crucial step in the application of a cluster analysis. Since the present study is explorative in nature and focuses on theory building rather than testing, a cognitive approach was chosen to define the clustering variables. While both inductive and cognitive techniques help to generate a rich description of the sample's characteristics, the latter technique is preferred as it captures the experiences of industry experts and thereby increases confidence that the variables are relevant and meaningful.[9]

In order to identify the grouping variables, interviews with nine brand-protection and production experts from the luxury goods, fast-moving consumer goods and aviation industries were conducted. Each practitioner was asked to name the five most important characteristics of counterfeit articles which allow conclusions on the strategies of counterfeit producers to be drawn. The five most often cited properties were:

- (v1) visual quality;
- (v2) functional quality;
- (v3) product complexity, where this attribute relates to the added complexity from counterfeit producers (i.e. adding a counterfeit label to a complex generic product only leads to a low score);
- (v4) potential loss or danger for the user; and
- (v5) degree of conflict with the law in the country of production.

---

[9] See, for example, Meyer et al. (1993) and Ketchen and Shook (1996). Application of the analysis can be found in Mascarenhas and Aaker (1989) and Reger and Huff (1993).

The respondents were confident that other industry experts could also provide reliable information on these variables as long as they were familiar with the product under study and the corresponding production techniques. A second set of attributes which requires a higher degree of interpretation on the part of the respondents was suggested:

- (v6) estimated investment in production facilities and organization;
- (v7) estimated specialization regarding product and brand selection; and
- (v8) estimated output given the applied production technology.

The measurement scale for each variable was defined in a group discussion to ensure comparability among the different analyses. Table 2.1 summarizes the suggested scales for each dimension. Further analysis showed that the more concise first set of variables was sufficient to identify meaningful and distinct groupings of the observations. Variables v2 and v4 were highly correlated, so v4 was excluded

Table 2.1: The scale of the variables

v1: Visual quality	v5: Degree of conflict with law
1 = Counterfeit origin obvious for non-expert without closer inspection	1 = Tolerated by authorities in country of production
2 = Counterfeit origin obvious for non-expert only after closer inspection	3 = Tolerated with some connections to enforcement agencies
3 = Counterfeit can be recognized by suspicious consumer only after closer inspection	3 = Tolerated only with very good connections
4 = Difficult to distinguish for product expert	4 = Risk of considerable punishment
5 = Difficult to distinguish for counterfeit expert	5 = Considerable risk of life time imprisonment or death penalty
v2: Functional quality	v6: Investment in production facilities and organization
1 = Counterfeit has no functionality / effect	1 = Less than USD 5,000
2 = Very limited functionality for a short time	2 = USD 5,000 to USD 50,000
3 = Functional quality considerably lower than of a genuine low-cost alternative	3 = USD 50,000 to USD 500,000
4 = Funct. comparable to genuine low-cost product	4 = USD 500,000 to USD 5,000,000
5 = Functionality equal to generic product	5 = USD 5,000,000 or more
v3: Product complexity	v7: Specialization
1 = Only label attached	1 = Product and brand can be changed at low cost
2 = T-Shirt, belt	2 = Product category can be changed at low cost
3 = Quality handbag	3 = Product category fixed, brand can be changed at low cost
4 = Medium to high quality mechanical watch, hand mixer, simple combustion engine	4 = Highly cost intense to change product
5 = TV and more complex products	5 = Highly cost intense to change brand or product
v4: Potential loss or danger for user	v8: Output
1 = No significant financial loss	1 = Less than 0.1 percent of licit production
2 = Some financial loss (USD 10 to USD 1000)	2 = Less than 10 percent of licit production
3 = Considerable financial loss (over USD 100)	3 = Less than 33 percent of licit production
4 = Threat to health and safety (for example allergic reaction, bruises, burns)	4 = Less than 100 percent of licit production
5 = Potential deadly injuries	5 = Output exceeds licit production capacity

from the further analysis.[10] While the analysis based on all eight variables led to the same principle groupings, the solution with a smaller number of variables (v1, v2, v3, v5) was preferred, following Punj and Stewart's (1983) recommendation for a small number of group-defining variables in a cluster analysis. Table 2.2 provides the descriptive statistics and the Pearson correlations of the group-defining measures.

Table 2.2: The Pearson correlations and descriptive statistics for sample

Variables	Means	Std. dev.	1	2	3	4	5
1. Visual quality	3.45	0.95		0.38*	0.30*	−0.07	0.02
2. Functional quality	2.55	1.13			0.53*	−0.71**	−0.41*
3. Product complexity	1.84	0.70				−0.32*	−0.43*
4. Potential loss or danger	2.70	1.24					0.53*
5. Conflict with the law	2.96	1.45					

N = 119; *correlation is significant at the .01 level; **correlation is significant at the .005 level

Data sample and sample characteristics

The sample was gathered over a 20-month period. In order to ensure a broad sample base, at least ten counterfeit cases from each of the following product categories were selected for further investigations: (1) foodstuffs, alcoholic and other drinks; (2) perfumes and cosmetics; (3) clothing and accessories; (4) electrical equipment; (5) computer equipment (hardware); (6) watches and jewelry; (7) cigarettes; (8) pharmaceutical products; (9) mechanical parts; and (10) fast-moving consumer goods. Overall, these ten categories make up over 80% of all counterfeit cases as reported by European customs (c.f. TAXUD 2005, breakdown by product category). Each category is represented by at least one, mostly two, brand owners with a market share among the top ten within their market segment. Interview partners within these companies were identified through telephone calls or were already known to the authors from prior research projects. The interviews were announced by pre-calls and an introductory email. The willingness to support this project was high: 25% of the contacted enterprises provided the required information, 65% refused to participate mainly due to concerns over talking about delicate issues with people from outside the firm (60%) or without giving any specific reason (30%), and 10% claimed they did not have any samples of counterfeit articles at hand. The high response rate may have resulted from earlier joint research cooperation

---

[10] This is done since the information contained in v4 is sufficiently captured by v2.

among many of the targeted companies and our institute; moreover the majority of companies signaled considerable interest in the results of the study, which the researcher promised to make available to them after the conclusion of the project.

Overall, the characteristics of 128 articles comprising 38 brands were discussed with brand-protection and manufacturing experts. The experts were asked to choose articles from the most recent three to six counterfeit cases their company had been involved in, in order to prevent a selection bias towards extraordinary or particularly spectacular cases. Samples of all but 34 articles were physically available during the interviews so as to reduce errors resulting from bad memory on the

Table 2.3: Industry categories represented in the sample

Product category	Category description	Cases in current	sample
(i)	Foodstuffs, alcoholic and other drinks	11	9 %
(ii)	Perfumes and cosmetics	13	10 %
(iii)	Clothing and accessories	20	16 %
(vi)	Electrical equipment	10	8 %
(v)	Computer equipment (hardware)	9	7 %
(vi)	Watches and jewelry	13	10 %
(vii)	Cigarettes	10	8 %
(viii)	Pharmaceutical products	11	9 %
(xi)	Mechanical parts	16	13 %
(x)	Fast-moving consumer goods	11	9 %
	Other goods	3	2 %

Info Box 2.1: What is cluster analysis?

Cluster analysis is a mathematical method to partition a data set (e.g. on people, companies, things, chemical processes, etc.) into subsets (clusters), so that within-subset-variation is small and between-subset-variation is large. In other words, cluster analysis (ideally) helps to define different groups so that each group has mostly similar members, but the groups are different from each other. Such categorizations can help to reveal structures within large amounts of data, ease an interpretation of the observations, and thereby facilitate further analyses. Data clustering is a common technique used in data mining, speech recognition, image analysis, and bioinformatics. It is also frequently used in business research, e.g. when assigning many different customers to a small number of groups that can be more effectively targeted with specific marketing messages, when selecting geographic markets, or when classifying competitors. For an in-depth discussion of this technique, see Aldenderfer and Blashfield (1984). We will use cluster analysis to show that a number of different, characteristic types of counterfeit producers exist and to highlight their basic characteristics.

part of the respondent, and to limit the influence of perceived expectations of the interviewers. Table 2.3 provides an overview of the counterfeit articles and the corresponding product categories.

Results and test for reliability and validity

Cluster analysis was applied to the above-mentioned data set. The analysis revealed that a number of combinations of specific parameter values (i.e. specific individual characteristics of counterfeit producers) appear to be much more likely than others, allowing for a segmentation of the analyzed articles. In fact, five clearly distinct groups of counterfeit producers can be identified:

Group 1 produces counterfeit goods with the lowest average visual quality. The average functional quality was rated as medium, in most cases allowing the owner to use the product, but having to sacrifice durability, stability, performance or contingency reserves. The typical product complexity is low to medium and a further analysis showed that many producers within this category target branded articles with high interpersonal values. The expected conflict with law enforcement in the country of production is the lowest among all the groups. Since members within group 1 primarily utilize the disaggregation between brand and product, they can be labeled Disaggregators.

Group 2 produces counterfeit articles with the highest visual and functional quality. Product complexity was highest among all groups, often allowing for an actual consumption or usage of the counterfeit articles. Counterfeit actors within this category seem to face limited pressure from local enforcement agencies. Since the product-related characteristics of the members within group 2 are most similar to those of the genuine articles, this group can be referred to as Imitators.

Group 3 is made up of producers of articles with a high visual, but low functional quality. Products are typically of medium complexity and are likely to pass as genuine articles if not carefully examined. They may result in a substantial financial loss for the buyer or even endanger the user's health and safety. Consequently their producers often face considerable punishment if their activities become known. Since the deceptive behavior towards the buyer of the corresponding article constitutes the main characteristic of the producer, this group of counterfeiters can be labeled Fraudsters.

Group 4 contains producers of goods of medium to high visual quality, but with the lowest functional quality and product complexity. Products within this category are likely to severely endanger their user or consumer. Consequently their producers potentially face extensive conflicts with enforcement agencies. For further discussion actors within this group are termed Desperados, pointing out their unscrupulous behavior.

Group 5 is made up of producers of articles with an average high visual and functional quality and a medium complexity. In this respect group 5 resembles group 2. However, the expected conflict with law enforcement agencies is significantly higher since most actors within group 5 target branded products upon which the state imposes high taxes. Group 5 can be referred to as Counterfeit Smugglers since they primarily profit from circumventing taxes rather than from realizing brand-name-related earnings.

Table 2.4 (a) provides the quantitative results of the group defining for each identified cluster.[11] Table 2.5 shows the distance between the cluster centers. A thorough interpretation of the groups and a discussion of the likely strategies of the actors will be provided later in this section.

Table 2.6 shows the cross-tabulation of the producers' group memberships and the targeted product categories. Based on the results of Fisher's exact test, the null hypothesis that the counterfeiters' strategy types are randomly distributed across the product categories is rejected. In fact, certain strategy types are predestinated for certain counterfeit goods. Counterfeit Smugglers, for example, in addition to the brand-name-related earnings, rely on making profits by evading taxes and therefore are likely to concentrate on bootleg tobacco products and alcoholic beverages. The quality of pharmaceutical products is especially difficult to assess prior to the purchase, making this category attractive for Desperados. However, the strategy types seem to be dependent on, but not merely surrogates of, the counterfeit product categories. Perfumes and cosmetics, clothing and accessories, electrical equipment, mechanical parts and fast-moving consumer goods are manufactured by at least three types of counterfeiters. Therefore the correlation between the product category and membership in a strategic group is not found to impose restrictions on the explanatory power of the study.

Throughout the study within-method triangulation served as an important tool to ensure reliability. The convergence of the results obtained by different clustering algorithms and distance measures indicates a high consistency of the solution, which in turn is an indicator for the reliability of the results (c.f. Hair et al. 2005). Furthermore, a split sample procedure, where the data set was randomly divided into half, was applied. Clustering of the subset also leads to a grouping with five distinct clusters with center means almost identical to those of the clustering results of the entire data set (c.f. Table 2.7). This again indicates a high degree of reliability in the findings.

---

[11] The results from the F- and Tukey Test were included, but only for descriptive purposes since the groups were chosen to maximize the differences among cases in different clusters.

Table 2.4: Statistics of group-defining (a) and non group-defining variables (b)

Variables (a)	Group 1: Disaggregators	Group 2: Imitators	Group 3: Fraudsters	Group 4: Desperados	Group 5: Cf. Smugglers	F (b)	Tukey test (c)
(a) Group-defining variables							
Visual quality	2.43 (0.79)	4.3 (0.60)	3.45 (0.74)	3.13 (0.76)	3.86 (0.66)	24.83*	2,3,4,5 > 1; 2 > 3,4; 5 > 4;
Functional quality	2.52 (0.51)	3.77 (0.43)	1.96 (0.63)	1.04 (0.21)	3.64 (0.50)	133.86*	2,5 > 1; 1,3,5 > 4; 2 > 3,4; 5 > 3;
Product complexity	1.83 (0.58)	2.47 (0.73)	1.83 (0.38)	1.13 (0.34)	1.71 (0.61)	19.59*	2 > 1,3,4,5; 1,3 > 4; 4 > 5;
Conflict with the law	1.39 (0.50)	1.83 (0.46)	3.00 (0.53)	4.83 (0.39)	4.79 (0.43)	249.43*	2,3,4,5 > 1; 3,4,5 > 2; 4,5 > 3;
(b) Non-group-defining variables							
Estimated investment	2.00 (0.67)	3.07 (0.69)	2.14 (0.64)	1.22 (0.42)	3.50 (0.76)	41.16*	1,3 > 4; 2,5 > 1,3,4;
Estimated specialization	2.61 (0.78)	3.47 (0.51)	2.76 (0.64)	1.78 (0.74)	3.56 (0.63)	24.45*	1,3 > 4; 2,5 > 1,3,4;
Estimated output	2.83 (1.03)	3.33 (0.80)	2.62 (0.82)	2.00 (0.63)	3.50 (0.76)	11.58*	1 > 4; 2,5 > 3,4;
Potential loss or danger	1.83 0.83	2.07 0.87	3.24 0.58	4.35 0.78	1.64 0.63	51.01*	1,3,5 > 4; 2 > 1,3,4,5;
Number of items	23	30	29	23	14		

(a) Means are shown, with standard deviations given in parentheses.
(b) Degrees of freedom for all variables are 4, 114.
(c) Groups are significantly different (p < .005) for Tukey's test in multiple comparison of means.

$* \ p < .001$

Table 2.5: Distances between cluster centers

Cluster	Disaggregators	Imitators	Fraudsters	Desperados
Imitators	2.37			
Fraudsters	1.98	2.40		
Desperados	3.87	4.42	2.18	
Cf. Smugglers	3.85	3.08	2.49	2.76

Table 2.6: Product type vs. group membership

Product category	Group 1	Group 2	Group 3	Group 4	Group 5
(i) Foodstuffs, alcoholic and other drinks	0	0	5	2	4
(ii) Perfumes and cosmetics	2	2	8	0	0
(iii) Clothing and accessories	10	8	2	0	0
(vi) Electrical equipment	1	3	3	2	0
(v) Computer equipment (hardware)	0	4	4	0	0
(vi) Watches and jewelry	6	4	0	0	0
(vii) Cigarettes	0	0	0	0	10
(viii) Pharmaceutical products	0	0	0	10	0
(xi) Mechanical parts	1	3	3	9	0
(x) Fast-moving consumer goods	2	5	4	0	0
Other goods	1	1	0	0	0

(Column group labels, shown rotated: Group 1 = Disaggregators, Group 2 = Imitators, Group 3 = Fraudsters, Group 4 = Desperados, Group 5 = Cf. Smugglers)

Table 2.7: Group characteristics of the holdout sample

Group-defining variables	Group 1: Disaggregators	Group 2: Imitators	Group 3: Fraudsters	Group 4: Desperados	Group 5: Cf. Smugglers	F (b)
Visual quality (a)	2.46 (0.60)	4.26 (0.32)	3.23 (0.36)	3.11 (0.36)	4.00 (0.00)	19.08*
Functional quality	2.54 (0.27)	3.68 (0.34)	2.31 (0.40)	1.01 (0.00)	4.00 (0.00)	52.72*
Product complexity	1.85 (0.14)	2.63 (0.58)	1.77 (0.19)	1.11 (0.11)	1.80 (0.70)	11.79*
Conflict with the law	1.54 (0.27)	1.95 (0.16)	3.15 (0.14)	4.78 (0.19)	4.80 (0.20)	120.57*
Number of items	13	19	13	9	5	

(a) Means are shown, with standard deviations given in parentheses.
(b) Degrees of freedom for all variables are 4, 54.   * $p < .001$

Careful validation is essential to assure that a meaningful and useful grouping of observations is arrived at. In this context reliability as demonstrated above is a necessary, but not a sufficient condition (Kerlinger and Lee 1999). In order to assess the validity of the findings, the insights and experiences of external practitioners can allow for a between-method triangulation as their perspectives are likely to differ from the researchers' expectations and judgments (Ketchen and Shook 1996). Therefore external expert knowledge, i.e. industry-affiliated brand-protection and anti-counterfeiting specialists other than those who helped to identify and select the group-defining variables, was taken into account to validate the findings. The respondents found the five-cluster solution clearly reflected the supply-side of the counterfeit market. Again most industry experts were able to provide a consistent analytical interpretation of the results without any prior explanation from the research team.

In fact, the small number of variables led to the confined clusters and the clusters' members show common characteristics beyond those that result directly from the clustering variables. Table 2.4 (b) summarizes the characteristics of the non-group-defining variables for each cluster and provides the corresponding results of the Tukey Test and t-Test. Moreover, the experts' opinion underpinned the practical value of the study. In fact, the results support hypothesis generation for further studies as well as the development of management recommendations. The implications are outlined below.

Main findings

The previous analysis supports the existence of five distinct types of counterfeit producers, each with different production capabilities, different foci on visual and functional quality (i.e. different emphasis on the consumers' pre-purchase and post-purchase experiences), and different associated risks with respect to prosecution. This positioning, be it due to external constraints or due to a deliberate choice, can be interpreted as the strategy of individual counterfeit producers. In the following discussion the existence of the strategic types is substantiated by presenting additional characteristics of corresponding counterfeit producers and by providing reasoning for the formation of each group. The findings result from semi-structured interviews and group discussions with brand-protection experts from industry and enforcement officers from customs, and were also validated against internal documentation of counterfeit cases, including raids, confiscations of stocks and seizures of production machinery.

Disaggregators focus on producing products with an average[12] functional quality. The potential financial loss or danger for the user is typically low as are the expected conflicts with local law enforcement agencies. Targeted product categories are mostly clothing and accessories, as well as luxury consumer goods with high interpersonal values, though the activities of Disaggregators are not limited to these categories. The business case seems to build upon generating brand-name-related earnings with minimal investments in production facilities. Trademarks either enrich generic goods or substandard products merely serve as a carrier for a

---

[12] Here, the term 'average' relates to the quality of counterfeit merchandize.

trademark (for example one type of handbag which is available with various labels). This statement is also supported by the low to average complexity of the counterfeit articles. A low investment in machinery and facilities limits the financial loss in case of raids, but also confines production to easy-to-manufacture goods. These products are very often of inferior quality and must sell as non-deceptive counterfeits for a fraction of the original product's price. As low sales prices do not justify expensive shipment strategies or direct selling, counterfeiters use large consignments to export their goods and rely on middle men in the country of destination in order to supply street vendors. This not only reduces the margin of the illicit manufacturer, but also makes the products susceptible to seizures. As a considerable part of the illicit value chain is in the country of sale (because distribution is risky and cumbersome), the extent of the corresponding counterfeit articles greatly depends on the efficiency of the enforcement activities in this country.

Imitators produce counterfeit articles with a relatively high visual and functional quality. An analysis of the non-group-defining variables consistently revealed a high average estimated investment, a high degree of specialization, and a relatively high production output. In many cases the corresponding counterfeits fulfill the needs of the user, but the functional quality is clearly below that of the corresponding genuine products. An important finding from the interviews was that Imitators often primarily serve their home market. In young economies where intellectual property rights are not strictly enforced, the use of foreign patents and designs can help companies to reduce their efforts during development processes, and significantly lower the risks of product launches. Similarly, trademark infringements can foster sales, thus establishing economies of scale and accelerating experience curve effects. Counterfeiters within this group are most likely to turn into licit competitors once intellectual property rights become more strictly enforced.

Fraudsters typically produce articles of a high visual, but low functional quality and aim to sell these goods as deceptive counterfeits. They often target products where the buyer is likely to be unaware of the existence of faked articles (prominent examples are fast-moving consumer goods). This enables Fraudsters to realize sales prices close to those of genuine products, thus justifying losses due to eventual seizures. These characteristics seem to be reflected in a low estimated investment in production facilities, which can be interpreted as an attempt to preserve flexibility and to limit the financial loss in the case of seizures of equipment. Interviews with brand-protection experts also revealed that Fraudsters often aim to infiltrate the supply chain of licit companies.

Desperados have similar characteristics to Fraudsters, but they take a more extreme position with respect to endangering the well-being of the end-consumers. They mostly target expensive, but simple-to-mimic products, such as pharmaceuticals or automotive spare parts, whose quality is difficult to evaluate prior to purchase. Desperados face the risk of severe punishment. However, this is to be seen alongside the considerable profits, even if most of the fakes were confiscated. To reduce their risk, Desperados mostly produce on a small scale, which is also reflected in the low score of the non-group-defining variable "Estimated investment in production facilities".

Counterfeit Smugglers have a special position as they primarily realize profits by evading taxes rather than by gaining brand-related earnings. Prominent examples are alcohol and tobacco products. High profits are juxtaposed with stringent actions by government agencies. A common characteristic of the members within this group is their strong ties to organized crime, a high level of investment in the protection of their operations, and, partly related to the latter, a high degree of vertical integration from production to distribution.

We will greatly benefit from this classification when discussing the development and selection of suitable protection and prevention strategies.

2.2   *Distribution channels and shipment strategies for illicit goods*

Counterfeit producers often rely on elaborate logistics skills that allow them to distribute their products while concealing their illicit nature, disguising the location of their production plants and protecting intermediate stakeholders. They go to great lengths to try and ensure their goods are not confiscated as seizures not only reduce their profits, but also increase the risk of backtracking individual shipments. The latter eventually jeopardizes the counterfeiters' sales channels, which are often expensive to establish and may ultimately lead to a confiscation of production machinery and prosecution of the actors. Consequently an understanding of the flow of counterfeit goods can help licit manufacturers, brand owners, and enforcement agencies to protect the licit supply or even to disrupt counterfeit activities. In this context a model is introduced to illustrate the flow of goods between counterfeit producers and licit actors. The model aims to help practitioners to systemize their supply chain security efforts, and thereby emphasizes the notion of counterfeit trade as an organizational, industry-like phenomenon.

The illicit supply chain

Companies can make or buy inputs, transfer outputs downstream or sell them. Illicit actors can do the same. In fact, counterfeit goods exist in final and intermediate markets. In order to protect licit companies from counterfeit goods infiltrating their supply, managers have to eliminate the value chain's permeability to such goods. This, however, requires some knowledge concerning the structure of the illicit market and the way it interfaces with the licit supply chain. In an attempt to structure an analysis, a visual representation of the licit and illicit flow of goods is given in

Figure 2.1. Goods move from the part supplier to the component supplier, the manufacturer, the distributor, and the retailer, who then passes them down to the consumer. The final link in the chain is waste management. Individual steps may be omitted and additional steps may be introduced to better reflect the supply chain of the company or industry under study. Several intermediate markets are included, allowing for easy integration of additional stakeholders.

The illicit supply is represented analogously. Goods pass through the chain as non-deceptive counterfeits and are sold to the licit side via intermediate or final markets as deceptive counterfeits. Though observations of the counterfeit markets suggest that illicit actors exist at all steps of the value chain, it is likely that several steps can be neglected in company or industry-specific settings; counterfeit manufacturers may, for example, buy their supply from the licit market instead of relying on illicit parts suppliers, and serve their retailers directly without intermediate markets and distributors.

After integrating other relevant stakeholders or eliminating obsolete steps from the value chain, an actual instantiation of the model can be developed where the individual stakeholders (boxes) and distribution paths (arcs) are described. Of interest are both typical properties and classification of boundaries (for example

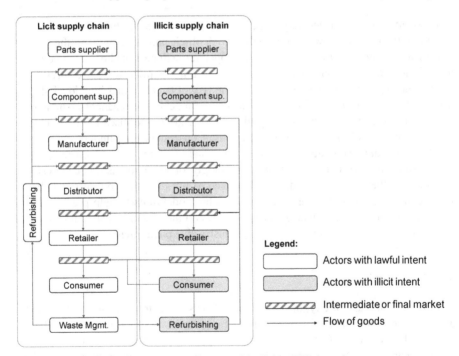

Figure 2.1: The coexistence of the licit and illicit market

unlikely behavior or characteristics). Especially the following issues should be addressed:

- *Characterization of the actors in each box.* Knowledge of the basic characteristics of counterfeit producers or distributors allows monitoring and prevention measures to be refined. Actors can be evaluated regarding their strategic focus or production settings as detailed earlier in this chapter. Relevant issues concern the location of production, appearance in the market, typical product characteristics and potential collaboration with other stakeholders.
- *Identification of frequently used paths and jeopardized intermediate markets.* Knowledge of transport routes and intermediate markets is important as it can help licit actors to disrupt counterfeit activities and to protect their own business from the infiltration of imitation products. Recently used transport routes and markets (including online markets) have to be identified and evaluated, and likely contingency routes of illicit actors should be pointed out.
- *Analysis of typical shipment strategies.* For the further instantiation of the model, users may turn their attention to the shipment strategies of illicit actors, which often include transshipments where illicit actors break their routes to disguise the origin of the goods, the use of small lot sizes which make seizure expensive, the use of expensive postal services which are rarely investigated, admixing original products with counterfeits to reduce the chance of detection even if the shipment is investigated, and the use of shipments which are similar to original shipments in terms of quantity, traffic route, and appearance.
- *Integration of customs.* Each flow of goods is a transaction potentially involving a border crossing. Customs is a major stakeholder in the battle against counterfeiting, but often relies on information from licit manufacturers or brand owners in order to recognize intellectual property rights infringements. Relevant customs checkpoints can be included in the model.

If a company is threatened by counterfeit products of different categories, several instances of the model may have to be defined. The same may be necessary for different geographic markets, especially with respect to non-deceptive counterfeit consumer articles, as the importance of various sales channels may vary. A survey conducted in Great Britain and Northern Ireland, for example, revealed that of those counterfeits which were knowingly purchased, 54% were bought outside the country, 28% on street markets, and 4% in domestic stores (Bryce and Rutter 2005). The same question directed to German-speaking consumers revealed a different market structure. Here, 78% of the non-deceptive counterfeit articles were bought abroad, 11% over the Internet, 8% at work or school, and less than 2% in shops or from street vendors.[13] Based on individual instantiation, companies can conduct a risk analysis and develop or adjust their anti-counterfeiting strategy accordingly.

---

[13] The survey was conducted between April and June 2005. 203 randomly chosen respondents over the age of 14 were asked where and in what quantities they had bought counterfeit goods within the previous year.

Info Box 2.2: Counterfeit cases by means of transport

A large proportion of counterfeit cases are initiated after phony products have been identified in air cargo freight and postal items. In 2006, these expensive means of transport accounted for more than 75% of all cases within the European Union. Shipments by mail are also very attractive for counterfeit actors as they do not require additional intermediate stakeholders who would have to dispatch and further distribute the goods. Their high share also reflects the importance of direct-selling over the Internet. Only 8% of the cases are initiated after inspections of sea freight. This, however, does not mean that sea fright is less susceptible to counterfeit trade. There, due to the larger lot sizes, individual cases often amount to thousands of illicit imitation products. Moreover, one may raise the question if sea containers are less thoroughly inspected than air cargo.

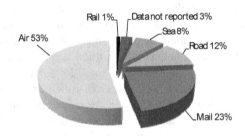

Source: (EC 2007)

Info Box 2.3: Number of articles seized by countries of origin

The table below shows the number of articles seized by European customs in 2006, expressed as percent by origin/provenance and product type. It provides a reasonable overview on where illicit imitation products were manufactured and also shows which countries were used as transshipment hubs to disguise the real origin of some goods.

China's dominance in counterfeit trade is well-known. The distance to her nearest illicit competitors, however, is nevertheless significant. The People's Republic is the biggest importer of counterfeit goods in all but two product categories, leaving India, Turkey, Ukraine, and the United Arab Emirates far behind. In fact, very few countries are responsible for the vast majority of counterfeit imports. This statement holds not only for imports to the European Union, but also to North America. Other countries that appear on the list have very likely been misused as "safe harbors" to disguise the shipments' real origin. Switzerland, for example, is certainly no vivid producer of counterfeit sportswear. Choosing the home of the United Nations' World Intellectual Property Organization (WIPO) as a hub could almost reveal a strange sense of humor among illicit actors. It nevertheless highlights that the country of origin as declared in the freight papers is no sufficient indicator for the goods' integrity.

Foodstuffs, alcoholic and other drinks	18% Turkey	1% China	12% Singapore	7% Hungary	6% Tunisia	2% Argentina	2% Ukraine	36% Others
Perfumes and cosmetics	37% China	19% Ukraine	17% Indonesia	9% Arab Emir.	4% Turkey	3% Hungary	2% Hong Kong	9% Others
Clothing and accessories	63% China	5% India	3% Turkey	2% Ukraine	2% Hong Kong	1% Vietnam	1% Bulgaria	6% Others
a) sportswear	43% China	13% Vietnam	7% Swiss	5% Turkey	4% Bulgaria	4% Arab Emir.	3% Hong Kong	22% Others
b) other clothing (ready to wear)	50% China	19% India	9% Turkey	4% Pakistan	2% Bulgaria	2% Thailand	2% Hong Kong	11% Others
c) clothing accessories (bags, sunglasses..)	81% China	2% Malaysia	2% Algeria	2% Egypt	2% Bulgaria	1% Arab Emir.	1% Italy	9% Others
Electrical equipment	61% China	21% Hong Kong	7% Arab Emir.	3% Rep. Korea	2% USA	1% Taiwan		2 % Others
Computer equipment (hardware)	47% China	17% Spain	15% Hong Kong	4% Pakistan	3% Singapore	3% Egypt	3% Taiwan	8% Others
CD (audio, games, software), DVD, cassettes	88% China	5% Iran	1% Taiwan	1% Syria				5% Others
Watches and jewellery	72% China	19% Hong Kong	2% Rep. Korea	1% Malaysia	1% Thailand			4 % Others
Toys and games	85% China	3% Hong Kong	2% Spain	2% Vietnam	1% Pakistan	1% Syria	1% Paraguay	2% Others
Cigarettes	83% China	6% Arab Emir.	2% Algeria	2% Egypt	1% Serbia	1% Malaysia		5% Others
Medicines	31% India	31% Arab Emir.	20% China	4% Thailand	2% Philippines	1% Vietnam	1% Jordan	10% Others
Other	82% China	3% Turkey	2% Hong Kong	2% Bulgaria	1% Netherlands	1% Egypt	1% USA	8% Others

Source: (EC 2007)

The importance of customs in the fight against counterfeit trade

Customs play a key role in the fight against counterfeiting and piracy. Of all the articles seized in Europe, more than 70% of counterfeit goods are intercepted by the authorities. The TRIPS Agreement confers an essential responsibility upon customs, especially in an international context, and the location along national borders, the detailed knowledge of international trading routes, as well as the right to inspect the goods under their control enables the authority to exercise its duties as an effective gatekeeper. When customs officers have sufficient grounds for suspecting infringements of intellectual property rights, they may detain the goods for three working days, even before an application is lodged by the right holder (so-called ex-officio procedures), and ask the right holder to provide information on the case. Such IPR infringements include violation of trademarks, copyrights or related rights, patents, supplementary protection certificates (plant protection and medicinal products), designations of origin, or geographical indications. However, customs need the help of the right holders themselves to achieve significant results in the fight against counterfeiting and piracy. Applications for action provide customs with the information that helps to identify illicit goods and endow the authority with additional power for dealing with the suspected products. To submit an application for action, the right holder (or their representative) must fulfill two conditions. As stated on the customs website,

- "the application must provide customs with an accurate description to make identification possible" and
- "proof must be provided that confirms that the applicant is the holder of the right in question" (EC 2008).

Though the application for action is national in character it can, if deposited in a member state of the European Union, have the same legal status throughout the other member states. In such a case, a community application should be used (c.f. EC 2008). A model form of the community application is published in Official Journal of the European Union L 261 of 6/10/2007. Notes on its completion and the declaration form that the right holder must fill in are published in Official Journal of the European Union L 328 of 31/10/2004. After the filing of an application, customs will automatically seize the stopped goods, and the owner has ten working days instead of three to decide whether to take action. In 2006, more than three fourths of all seizures were initiated by applications for actions, and only 15% result from ex-officio interventions (EC 2007). Therefore, filing an application is strongly recommended.

2.3   Exploring the counterfeiters' business case

Given the industry-like characteristics of many production sites, cost-benefit calculations are almost certainly undertaken when selecting, changing, expanding or discontinuing illicit activities. For a better understanding of the underlying business concepts it is worth discussing the cost drivers of illicit production from a counterfeiter's perspective. We will furthermore contrast the potential financial advantages of such activities with the additional costs, risks, and limiting factors of future growth.

Counterfeit producers often reap the benefits that result from considerable investments by the corresponding brand owners. Obvious gains stem from free-rider effects with respect to research and development costs and marketing expenditure. Other significant savings result from the choice of raw material and production techniques that do not necessarily need to be up to the quality and safety standards of the corresponding brand owners. Moreover, counterfeiters have lower costs for salaries, taxes (if paid at all), reserve for warranties, compliance to environmental regulations, etc.; they neither face the risk associated to product launches and market entries nor do they have to offer complementary low margin products or services as they can concentrate on selected top-selling, well-established brands and products.

However, counterfeit production also comes with costs. These may include, for example, expenses for purchasing and maintaining production machinery, and the costs of raw materials and salaries. Moreover, counterfeit actors have expenses that licit companies do not have, for example for maintaining their illicit distribution network and dealing with local or government officials (which may require paying bribes or profit sharing). Other direct or indirect expenses include the cost ascribed to potential seizures of articles and production machinery as well as the risk of personal fines, imprisonment or even capital punishment. These numerous additional risks also increase the cost of capital for financing counterfeit activities and thus decrease the overall discounted profit of related investments.

The individual strategic setting of an illicit actor has a considerable influence on the cost of production and distribution, the sales prices and the scalability of the activities and long-term perspectives. Table 2.8 summarizes the most important cost factors and compares them with the expected revenue (i.e. sales prices and output). Again, each business model follows specific cost-benefit patterns and has specific advantages or disadvantages with respect to scalability and the venture's future perspectives. Low costs for reverse engineering, production equipment, raw materials etc. are put into perspective by considerably discounted sales prices (for Disaggregators) or by a high risk of prosecution (for Fraudsters and especially Desperados). Imitators may be able to realize long-term revenue streams but face comparably high expenses for reverse engineering, production and raw materials.

Table 2.8: Costs, revenue, and future perspectives of counterfeit producers

	Disaggregators		Imitators		Fraudsters		Desperados		Cf. Smugglers	
Product costs of illicit actors										
Reverse engineering	l		h		n		n		n	
Development costs	l	--	h	--	l	--	l	--	n	
Production equipment	l	--	h	-	l	--	l	--	m	-
Raw materials	m	-	h	-	l	--	l	--	m	o
Salaries	m	-	m	-	l	--	l	--	l	
Quality mgmt.	l	--	m	-	n		n	--	l	
Marketing	n	--	n	--	m	-	m	-	n	
Shipment of articles	m	+	m	+	l	+	l	++	m	+
Warranties	n	--	l	--	n	--	n		n	
Maintaining illicit distribution network	m		l		m		l		h	
Product seizures	m		h		m		l		l-m	
Confiscation of production equipment	l		h		m		l		m	
Bribes	l		l-m		l				h	
Fines, punishments	l		l		h		h		h	
Output										
Sales price	l	---	m	-	m-h	o	m-h	o	m	-
Output capacity	m		m-h		m		l		m	
Scalability of illicit production	l		h		m		l		m	
Long term growth	l		h		l		l			

Legend: Blank field: not applicable / no data

White field: absolute costs	o: none	l: low	m: medium	h: high
Gray field: comparison to licit producers	++ much higher	+ higher	o almost equal	- lower	-- much lower

Table 2.8 shows that seizures constitute a major cost driver for all but one strategic setting; they lead to a direct financial loss among illicit actors of at least the costs of production and transportation up to the point where the goods are confiscated. In general, losses are higher the more sophisticated the imitation products are. Moreover, indirect costs arise, for example when alternative transport networks have to be established. Counterfeit actors may need to search for new shipment routes, identify new logistics partners and establish new middle men, for example to bypass border controls. Not only are these activities laborious (particularly with regard to the limited transparency and the high search cost within such markets), they also increase the risk of running across unreliable partners or even informants.

As seizures increase the risk of prosecutions, higher margins may be demanded by the participating stake holders. Furthermore, questions on who to blame for the financial losses are likely to cause friction among the actors.

In fact, direct and indirect costs constitute a limiting factor for the growth of counterfeit trade, and product seizures contribute heavily to these costs. Figure 2.2 shows the axiomatic relationship between seizure rates and the resulting sales prices for licit and illicit actors for an exemplary good. The overall cost is approximated using Equation 2.1 which expresses the cost per successfully delivered product C as a function of the cost of production and delivery up to the point of destination $CoPD$, the seizure rate s, indirect costs due to seizures I, the desired margins of illicit actors M, and the per-article costs and margin of sales CaS. The model is exemplarily set up with the parameters as stated below.

$$C = \left(CoPD + CoPD\,\frac{s}{1-s} + I\,\frac{s}{1-s} \right) \cdot \left(1 + M_{\text{Production,Delivery}} \right) \cdot \left(1 + CaS_{\text{Retail}} \right) \qquad (2.1)$$

Cost (in units)	Licit actor	Illicit actor
R&D	1	0
Reverse engineering	0	0,01
Production facilities	8	1
Raw material	3	1
Labor	6	1
Logistics	4	8
Marketing	10	0
Warranty, quality mgmt.	1	0
Seizure (direct)	10% of seized goods' value	Dependent on seizure rate
Seizure (indirect)	0	Assumed to equal dir. costs
Margin manufacturer	30%	50%
Margin and cost retail	15%	30%
Taxes on margin	30%	0%

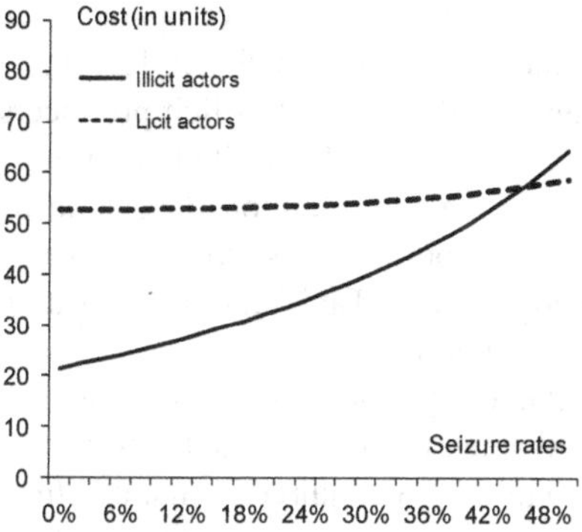

Figure 2.2: The impact of seizure rates on the cost of a counterfeiter's activities

Financing counterfeit activities

Industry-like production of goods requires professional financial management. Large-scale counterfeit producers (especially Imitators) have to raise capital to purchase machinery, facilities, etc., and have to maintain the ability to meet their financial obligations. Money lenders, as a matter of course, demand a return for their investments that not only includes a premium for the market risk but also for the risks associated with illicit activities and for dealing with criminal organizations. The lack of transparency of such activities and the difficulty to prevail over the "business partners" when trying to sell a stake restricts their access to finances even more. In fact, these aspects explain why other criminal organizations often stand behind large-scale producers of counterfeit goods. However, when counterfeit production is accepted by local officials or when the producers successfully disguise their counterfeit activities (for example by also producing non-counterfeits), capital may also come from licit sources (e.g. from investors who are not aware of the illicit activities).

Though the clandestine nature of the illicit market limits the accessibility to reliable data for example on the cost of capital in this domain, one should nevertheless reflect on the financial requirements of large-scale producers in order to fully understand the mechanisms of the counterfeit market. Approaches to limit the access to venture capital can also reduce counterfeit activities.

2.4   Research on counterfeit supply

Very few publications are dedicated to the supply-side issues of the counterfeit market, though knowledge in this field is of great importance for understanding the way the illicit market operates, how companies in emerging economies use imitation products to foster learning and development processes and how licit brand owners can fight illicit producers. One reason for the lack of related work is very likely to be the limited access to illicit market players and thus the difficulty of obtaining information on clandestine illicit market activities. However, some insightful publications exist, and we briefly summarize the most insightful contributions below. Table 2.9 provides an exhaustive overview of the academic literature in this field.

An early contribution on the supply-side was published by Harvey and Ronkainen (1985). The authors point out potential ways illicit actors may obtain the know-how required to manufacture counterfeit articles. However, their work is based mainly on the assumption that intellectual property is stolen from within the affected company, thus not reflecting the considerable reverse-engineering capabilities of today's counterfeit industry.

Olsen and Granzin (1992 and 1993) discuss how brand owners can prevent otherwise trustworthy distributors from knowingly or unknowingly selling illicit imitation products. The authors stress the importance of maintaining a high level of satisfaction and dependence among their supply chain partners, as well as showing their high commitment, in order to gain their assistance in fighting the counterfeit trade.

In an insightful case study Green and Smith (2002) detail the efforts of an international company to eliminate the production and distribution of counterfeit alcoholic beverages in an emerging Asian market. Thereby they also cover important characteristics of counterfeit production and detail the organizational structures of the illicit market. Their study provides evidence of a sophisticated production system characterized by a high degree of labor division and specialization, highly-protected individual operations organized in such a way that the elimination of a

single function or production site does not endanger other functions, and by strong
ties to organized crime. The case study also shows that, despite high margins from
the illegal activities, consistent seizures and raids have the potential to drive illicit
actors out of business.

Table 2.9: Supply-side investigations

Author(s)	Year	Short description
Harvey/ Ronkainen	1985	– Discussion of potential ways illicit actors can obtain classified information which enables them to produce counterfeit articles. – Loss estimates based on industry estimates.
Olsen/ Granzin	1992	– Depiction of how manufacturers can establish a relationship with their distributors to gain support in fighting illicit trade. – Interviews with five retailers from the automotive industry to conceptualize a structural equation model.
Olsen/ Granzin	1993	– Investigation of the influence of dependence, control, channel conflict and satisfaction on a dealer's willingness to help a manufacturer combat counterfeiting. – Findings are that manufacturers can engender cooperativeness by nurturing satisfaction and dependence in manufacturer-dealer relationships.
Glass/ Wood	1996	– Application of social exchange theory to investigate the influence of situational factors on the intentions to engage in software piracy. – Findings are relevant in the context of exchange in peer-to-peer networks but do not directly apply to commercial counterfeiting.
Green/ Smith	2002	– Summary on the literature that addresses counterfeit trade. – Strategies for addressing the threat in developing markets. – Case study of a major company producing and selling alcoholic beverages.
Ben-Shahar/ Assaf	2004	– Development of a formal model in which a manufacturer may promote copyright infringements to indirectly participate in predatory pricing and to deter competitors from entering the market.
Liu et al.	2005	– Effect of random examinations and different punishment levels with respect to store managers who potentially sell deceptive counterfeit products.
Khouja/ Smith	2007	– Analysis of profit maximization models, which take both piracy and saturation effects into account.

3 Counterfeit Demand and the Role of the Consumer

Consumers can take many different roles in counterfeit trade. They may buy counterfeit goods knowingly or in the belief that they are purchasing genuine products, they may try to ensure obtaining only original articles or invest considerable effort in acquiring less expensive fakes, and they can even become actively engaged in selling illicit products themselves. In fact, understanding their multifaceted roles is essential for evaluating the implications of counterfeit trade and for developing effective consumer information programs.

The following chapter aims to provide the essential insights into consumer behavior in markets where counterfeit products are available. We investigate the awareness and the willingness to purchase such goods, and we analyze the motives of those who intentionally buy fakes. The survey-based findings enable licit manufacturers to assess counterfeit-related risks for specific product categories and help to identify those buyers who are likely to intentionally purchase illicit goods, thus showing where relying on the consumers' help is or is not expedient. Furthermore, the investigation of consumers' reasoning for and against intentional purchases of fakes helps to find arguments to effectively influence public opinion on counterfeit trade. The empirical data also allows conclusions to be drawn on whether counterfeit consumers and consumers who do not purchase illicit goods form two distinct groups, or whether and to what extent both groups overlap, thus helping to develop a better understanding of alternative buying behavior and substitution effects.

3.1 Consumer behavior in counterfeit markets

Problem awareness, purchase intentions, demographic characteristics and the consumers' attitudes towards counterfeit trade are extremely important factors for the development of brand- and product-protection strategies. Capturing these demand-side characteristics, however, is difficult as consumers are often not aware of the counterfeit nature of a product or, for non-deceptive counterfeit cases, are unwilling to admit or explain their socially unacceptable behavior. In fact, developing an adequate survey design requires a great amount of sensitivity in order to precisely capture the reality of the illicit market. Relying on existing studies also calls for a good understanding of the underlying constructs (i.e. set of questions). Studies authored or commissioned by industry association, for example, often use biased constructs to highlight the relevance of the problem (c.f. Info Box 3.1). Some of their results may be extremely important for raising problem awareness but are

---

Info Box 3.1: Problems with existing counterfeit consumer surveys

The outcomes of consumer surveys very much depend on the way individual questions are formulated. With respect to counterfeit consumption, a careful design is especially important as questions on socially unaccepted, embarrassing, or unlawful behavior can reduce the willingness to participate in such studies and may lead to a bias towards answers that describe socially desirable behavior. Therefore investigators often try to compensate this bias by introducing additional premises. The question

"Which, if any, of the following goods would you knowingly purchase as counterfeit, assuming you thought the price and quality of the goods was acceptable?"

may serve as an example. The responses may indicate which products are particularly susceptible to non-deceptive counterfeit consumption. However, of what explanatory power are the answers from those participants who, outside the study environment, would never assume that the quality of whatsoever counterfeit good is acceptable? Therefore based on this question, conclusions such as "7% of the consumers would purchase counterfeit alcoholic beverages" are unlikely to reflect the real behavior. In fact, when interpreting such studies, there is no way around having a closer look at the individual questions and challenging their suitability for the consumer study.

---

not suited as a basis for decision making for management, while other findings are truly insightful if interpreted correctly.

In the following section we will refer to numerous demand-side studies designed to meet high academic standards. We will furthermore complement their findings using the results of our own surveys that we have conducted over a period of three years. Some of these surveys are – for the sake of higher reliability – brand- or product-specific. The reader should be aware that questions with respect to other brands and products may lead to different results. Guidelines for the design of studies focusing on other goods are given in Section 3.2.

Awareness and buying behavior

Product categories where a large proportion of consumers are not even aware of the existence of counterfeit goods are an easy and highly profitable target for illicit actors. Careful product inspection by the customer is less likely and even articles that are well below the expected level of quality rarely constitute a reason for distrust but are recognized as a failure of the brand owner. Moreover, illicit actors can sell their deceptive fakes at the original's price and thus are able realize high margins.

The awareness of counterfeit trade with respect to different product categories was investigated by the Anti-Counterfeiting Group (ACG) in a survey involving approximately 1,000 English-speaking consumers. The share of people responding

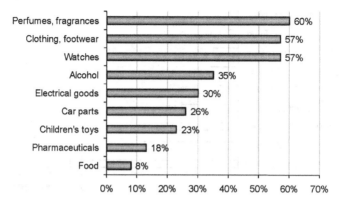

Figure 3.1: Awareness of counterfeit trade with respect to different products

with yes to the question "Are you aware of the sale of counterfeit goods in any of the following categories?" is shown in Figure 3.1 (ACG 2003). While the absolute numbers are questionable since "being aware" may denote various levels of consciousness and since it is difficult to say whether the respondent had been aware of the existence before he or she was reminded of it by the text in the questionnaire, the relative share is rather insightful.[14]

The findings reveal a considerable difference with respect to the various product categories under study. About 60% of the respondents were aware that counterfeits exist in categories which are often bought knowingly (for example clothing and watches) or that are at least offered on street markets, for example in countries that have been visited on holiday. However, consumers seem to be rather unsuspicious in categories where brands are seen as a sign of quality rather than as a means to communicate values and social status. Less than 30% of the respondents claimed to be aware of counterfeits with respect to foodstuffs, car parts, toys, pharmaceuticals, etc. Active engagement of the consumer in an authenticity test of those goods would therefore require some awareness training.

The same study also investigated the willingness to purchase different counterfeit products; the participants' positive answers to the question "Which, if any, of the following goods would you knowingly purchase as counterfeit, assuming you thought the price and quality of the goods was acceptable?" is shown in Figure 3.2. Though the validity of the result may be criticized as the question requires the participant to make an assumption which may differ from his or her natural statement towards the product, the results implicitly indicate that a potential health and safety risk strongly affected the purchasing decision. The findings are supported by the survey conducted by us in 2006 among 203 German-speaking respondents who were asked which counterfeit products they had knowingly bought during

---

[14] The absolute numbers when using "are you aware of " questions are even likely to be lower.

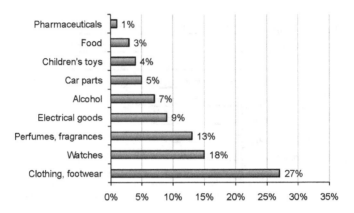

Figure 3.2: Willingness to buy counterfeit goods with respect to different products

the previous three years. While 33% said they had purchased counterfeit branded clothes, 27% handbags and fashion accessories, and 12% watches, only less than 2% said they had purchased counterfeit beverages, children's toys and pharmaceuticals. In the latter categories consumers are more likely to support the anti-counterfeiting efforts of brand owners.

Purchasing behavior with respect to exclusive brands

An important question with respect to counterfeits of exclusive branded products is whether and to what extent consumers who purchase counterfeits knowingly and those who do not engage in such activities form two different groups regarding their attitude and purchasing behavior of genuine goods. If a clear difference is observed, one can assume that counterfeit and genuine products are mostly sold into distinct markets, and direct substitution of one product by the other is limited. If, however, a considerable overlap between both groups exists, competition among counterfeit and genuine versions is more likely.

Purchasing behavior and attitude with respect to well-known, exclusive brands was tested based on established constructs (i.e. on a set of questions) that have been thoroughly tested in marketing science in order to ensure the highest possible level of validity of the findings.[15] Of interest were the perceived personal value (the utility a brand has for oneself without taking into account the opinion or thoughts of others), the perceived interpersonal value (the utility of a brand as a means to communicate, for example, wealth, social status, or membership to a group), as well as the perceived functional value and the quality associations (such as the expectations with respect to durability or precision). Two questions were used to cover each construct:

---

[15] We use many constructs that were introduced by Vigneron and Johnson (1999).

- *Perceived personal value.* "I can identify myself with certain exclusive brands" and "I also purchase such brands to reward myself for an achievement".
- *Perceived interpersonal value.* "Exclusive branded products are a sign of success" and "Exclusive branded products are a sign of good taste".
- *Perceived functional and quality-related associations.* "Products of well-known brands are often of better quality than no-name products" and "The design of products of well-known brands is excellent in most cases".

Two additional questions help to assess the attitude towards brands and general brand loyalty:

- *General attitude towards price premiums.* "The price premium of branded products compared to no-name products is mostly justified".
- *General brand loyalty.* "If I buy something, I often purchase products of well-known brands".

The actual survey was conducted in a waiting room in Zurich's main train station in August and September 2005.[16] Anonymity was ensured, but personal assistance was available even though it was only required by less than 5% of all participants. Approximately 25% of the potential candidates were willing to participate in the survey. 203 respondents filled out the questionnaire completely; the gender, age, and income distribution is shown in Figure 3.3. Due to the relatively small number of participants and the non-representative distribution of age among the respondents, the survey had the characteristics of an extended pretest, but nevertheless allows for a number of insightful conclusions to be drawn.

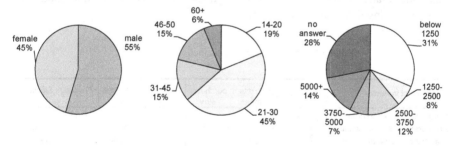

Figure 3.3: Demographics of the consumer survey: gender, age, and household income in EUR per month

---

[16] In Switzerland the public transportation system is used by people of almost all income classes and social classes. It was chosen because we were able to achieve a high response rate (compared to postal surveys) and avoided the bias that might have resulted from distributing the survey in shopping malls or streets that mostly fall in specific price categories.

Differences in attitude towards brands

The attitude towards brands among those who have purchased and those who have not purchased counterfeit goods in the last three years was based on measurements on the above-mentioned constructs. Possible choices on a five-point Likert Scale ranged from "strongly disagree" to "strongly agree". For the quantitative analyses, the choices were assigned corresponding integer values from −2 to 2.

First, an independent t-test was used to test for the difference between those consumers who had intentionally purchased counterfeit goods in the previous three years (n1 = 94) and those who had not (n2 = 109). As shown in Table 3.1, both groups significantly[17] differ only with respect to the belief that price premiums of brands are justified (higher affirmation among those who did not purchase fakes) and concerning their opinion that exclusive brands are a sign of success (higher affirmation among those who purchased fakes) (c.f. Figure 3.4). The differences between the groups' associations with respect to brands and quality, good taste, and superiority of the design were only minor. Moreover, the reported buying behavior with respect to genuine branded goods did appear to be highly similar among counterfeit consumers and the rest of the group.

The difference between both groups was also investigated using a discriminate analysis. Again, the dissimilarity was found to be rather limited. In fact, when trying to predict whether a consumer does or does not purchase counterfeit articles using the constructs to measure the attitude towards brands, the classification is only correct in 62% of the cases – that is only 12 percent points better compared to randomly assigning the respondents to groups. The findings provide strong evidence that many customers of counterfeit goods also consider genuine goods in their purchasing decision and vice versa.

Table 3.1: Brand attitude as a distinctive characteristic

Independent variable	No intent. purchase		Intentional purchase		F-value	Signifi.
	mean	stdv.	mean	stdv.		
I frequently buy well-known branded goods	0,35	1,15	0,38	1,05	0,03	0,871
Price premiums of brands are justified	0,18	1,07	-0,19	1,08	6,13	0,014
Branded products are of better quality	0,51	1,00	0,49	0,99	0,03	0,862
The design of branded products is superior	0,53	1,08	0,64	1,05	0,54	0,463
I can identify myself with branded goods	0,00	1,30	0,21	1,28	1,35	0,246
I purchase branded goods to reward myself	-0,63	1,40	-0,59	1,30	0,05	0,831
Exclusive brands are a sign of good taste	-0,50	1,24	-0,53	1,25	0,03	0,855
Exclusive brands are a sign of success	-0,62	1,22	-0,24	1,30	4,51	0,035

---

[17] "Significantly" relates to a 0.95 confidence interval. The results are not based on the assumption of equality of variance.

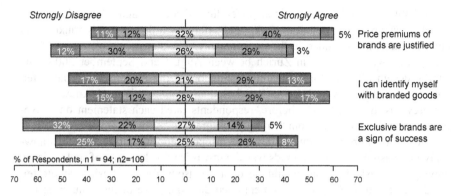

Figure 3.4: Attitudes towards brands; consumers who did not purchase (top) vs. consumers purchased counterfeit goods intentionally (bottom)

Reasoning for and against purchasing counterfeit goods

Questions to assess the reasoning for and against purchasing counterfeit luxury goods stemmed from several interview rounds in which the respondents were asked to mention five motives for each decision. Based on these interviews, a list of ten motives was compiled, which was then ranked. Scores were assigned in ascending order (most important given ten points, second most important given nine points, etc.). The five questions with the highest sum of scores were used in a test survey and later in the study outlined above.[18] Reasons for purchasing goods were:

- The good quality of counterfeits,
- the high price of the genuine article,
- the high value for money,
- the interest in counterfeits and the fun associated with having one, and
- the attractiveness of the brand and the unwillingness to pay for it.

Reasons against purchasing illicit goods were:

- The limited availability,
- the bad quality of fakes,
- the missing warranty,
- the better value for money of genuine articles in the long run,
- personal values, and
- potential conflicts with the law.

---

[18] Six potential reasons for a decision against non-deceptive counterfeit purchases were included due to an equality of points in the ranking process ("two fifth places").

Probably due to the focus on counterfeit luxury goods – and maybe also due to a lack of awareness – potential health and safety hazards were only ranked seventh, and thus not included in the questionnaire. The statements were part of our survey that we conducted in Zurich between August and September 2005. The survey design allowed for a separate evaluation of the responses of counterfeit consumers and those who have not purchased any counterfeit goods knowingly in the previous three years. As before, respondents rated each statement on a five-point Likert Scale labeled from "strongly disagree" to "strongly agree".

The findings are quite interesting. The primary reported motivation for knowingly purchasing counterfeit goods was the low price for the value of such articles. Having been asked "What would be or are reasons for purchasing counterfeit goods?", 65% of the respondents mentioned the high price of the genuine article as a strong or very strong reason, 58% the good quality of counterfeits, and 55% stressed the good cost-performance ratio of fakes. About 48% claimed that purchasing counterfeits for amusement ("just for the fun of it") would be a strong or very strong motivation.

Unlike the attitude towards brands, the reasoning for (potential) purchases differed significantly among buyers and non-buyers of counterfeit goods (c.f. Table 3.2 and Figure 3.5). The good quality of fakes, the high price of counterfeits, the attractiveness of brands, and the unwillingness to pay the genuine products' prices were found to be much stronger motives for those who had recently bought imitation products. The findings are in line with a survey conducted by Bryce and Rutter (2005), where cost as well as an "acceptable product quality" were the most frequently cited motivations for the purchase of counterfeit fashion items (72% and 60% of the respondents said so respectively).

Table 3.2: Reasons for purchasing counterfeit goods

Independent variable	No intent. purchase mean	stdv.	Intentional purchase mean	stdv.	F-value	Signifi.
The good quality of counterfeits	-0,07	1,63	0,72	1,43	13,84	0,000
The high price of genuine articles	0,35	1,72	1,05	1,33	10,51	0,001
The good value for the money of fakes	0,27	1,61	0,58	1,24	2,42	0,122
Buying counterfeit goods "just for fun"	-0,01	1,55	0,35	1,43	2,95	0,087
The attractiveness of the brand and the unwillingness to pay for it	-0,37	1,62	0,37	1,34	12,62	0,000

Primary reasons for not purchasing counterfeit goods were the poor quality of such articles and their limited availability (c.f. Figure 3.6); both motives were more pronounced among counterfeit consumers (c.f. Table 3.3). Interestingly, the groups assigned similar average scores to the statement "Originals are cheaper in the long run" (32% agreed or strongly agreed). As expected, the avoidance of counterfeits due to personal values was significantly more pronounced among those who had not engaged in counterfeit purchases (48% agreed or strongly agreed) than among the rest of the group (22% made the same statement).

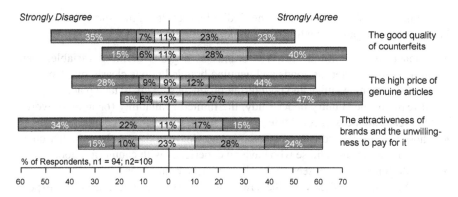

Figure 3.5: Reasons for buying counterfeit goods; consumers who did not purchase (top) vs. consumers who purchased counterfeit goods intentionally (bottom)

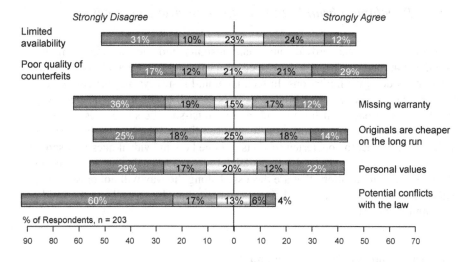

Figure 3.6: Reasons for not purchasing counterfeit goods

Table 3.3: Reasons for not purchasing counterfeit goods

Independent variable	No intent. purchase		Intentional purchase		F-value	Signifi.
	mean	stdv.	mean	stdv.		
The availability is limited	-0,51	1,45	0,07	1,46	8,11	0,005
The quality of counterfeits	0,02	1,50	0,64	1,44	9,03	0,003
The missing warranty	-0,51	1,58	-0,38	1,56	0,37	0,544
Genuine goods are cheaper in the long run	-0,24	1,42	-0,12	1,46	0,38	0,537
Due to personal values	0,33	1,54	-0,59	1,49	18,58	0,000
Due to personal values	-1,21	1,14	-1,18	1,31	0,03	0,867

Though we will refer to many other individual results of this survey later in the book, we may highlight the following important findings:

- When using the attitude towards brands as a group-defining variable, then markets for counterfeit goods and genuine branded goods clearly overlap; substitution among genuine and counterfeit products is likely to take place.
- At least for the country under study, the limited availability (or in other words the high search costs) constitutes a major factor that restricts counterfeit sales.
- Those who purchased counterfeit goods knowingly in the past have higher quality expectations than those who have no experience with counterfeit goods.
- Counterfeit consumers seem to be slightly more brand-prone.
- The potential conflict with the law that some counterfeit consumers expect appears to be only a weak reason for not purchasing illicit goods.

3.2   Developing brand- and product-specific consumer studies

Consumer surveys can provide important insights into the counterfeit market, provided that the questions are developed with care and that the analysts are aware of the methodology's limitations. In fact, meaningful surveys are difficult to compile, and the multitudes of product categories as well as the many different ways brands are positioned call for a brand- and product-specific survey design. In the remainder of this section we discuss the applicability and the limitations of survey studies with respect to counterfeit markets, outline how individual questions should be developed, and highlight the potential pitfalls when analyzing study results. The theoretical foundations that are discussed along the way will not only help the readers to design their own brand-specific surveys but also to evaluate the validity of commissioned studies.

What can be measured – and what not?

Questionnaires are a valuable tool to collect information from a large number of respondents. They allow something to be learned from the participants (for example when targeted to experts or experienced users), and for learning something about the participant (for example with respect to opinions, attitudes, demographics, etc.) – but always with the respondent as an intermediary. Though the last statements appear to be rather obvious, they lead to a number of important questions that have to be raised throughout the study design: Has the respondent the necessary capacity to understand the question and the knowledge to answer it? Is the knowledge accessible? Can he or she express it? Is the respondent willing to share it, and is he or she likely to be honest? Is there no hidden agenda, and if there is one, what are the respondent's intentions? Does the question provoke specific answers? Are generalizations of the findings plausible? These questions can clearly restrain the applicability of survey studies. Consumer experience with respect to deceptive counterfeit goods, for example, is difficult to access, as are experiences with goods that are rarely sold or the safety of counterfeit products (for example with respect to dangerous ingredients in toys). However, the reasons for or against purchasing illicit goods, personal opinions with respect to exclusiveness of brands and many other important factors that shape demand or allow for an evaluation of the impact of counterfeit trade can be made accessible.

Design process of the questionnaire – and aspects to think about during the evaluation process

Good construction of the questionnaire is critical to the success of the survey. Imprecise or inappropriate questions, wrong ordering of constructs, incorrect scaling, or a poor questionnaire layout can significantly reduce the value of a survey. The following guidelines shall aid the development process of a study as well as the evaluation of the validity of studies conducted by third parties.

If possible, use established constructs. Researchers spend much time to fine-tune individual constructs and used extensive test studies to evaluate the questions' validity and reliability. When compiling brand- and product-specific studies, one should definitely build upon the previous work as it dramatically increases the quality of one's study. The literature review in Section 3.3 helps to identify sources for tested constructs and survey designs.

If new questions must be developed, thoroughly check their reliability. The questionnaire should be composed of established constructs whenever possible. However, sometimes the required constructs are not available. In such cases it is important to come up with questions that are easy to understand (for the potential respondents!) and are clear without ambiguity. Questions may be developed in semi-structured interviews and selected in administered test surveys. Each question should be discussed with experts as well as with potential respondents to ensure that the items are correctly understood and that, in the case of closed questions, the set of possible answers is adequate.

Carefully put the questions in an adequate order. Use ice-breaker questions at the beginning (or at least avoid asking sensitive question at first), check for the influence of each question on the next construct, and use buffer questions to decouple individual constructs.

Ensure that the questions are easy to understand. The language used in the questionnaire must be adequate for the target group. Though one may think that this is an obvious issue, it is always worth checking twice whether it is understandable.

Furthermore, too many branches should be avoided in paper-based surveys to ensure an easy flow. The questionnaire should create a conversation-like feeling.

Take special care with displeasing questions. Asking displeasing questions may require some special techniques, even if some of them may water down the original aspect or first may look as if they were directed at something else. Techniques include projected questions ("Can you imagine why somebody would purchase counterfeit toys" rather than ("Why do you purchase…"), or, though a bit dodgy, questions that use limited imputations ("When did you purchase counterfeit products for the first time" rather than "Have you ever…"). In general it is more promising to go from unspecific to more specific questions to avoid blocking by the respondent.

In general, one should avoid questions that:

- *Deal with events that lie in the future.* Though future behavior may be of great interest, respondents are more likely to answer with what they think they should do. Where the answer may not be fully socially acceptable, it may be better to ask what a person did in the past and assume that the behavior remains unchanged.
- *Are suggestive.* When dealing with delicate or embarrassing issues (and counterfeit trade is perceived as such by many respondents), one should take great care to avoid normative questions as they greatly influence the choice of answers.
- *Are unbalanced.* When using questions with a predefined set of answers, the answer space should be symmetric. A possible set such as "disagree", "agree", and "totally agree" inevitably leads to a bias towards agreement. Though this technique is often used when the interviewer has a hidden agenda, is does not help to obtain valid market insights.
- *Contain more than one complete statement.* If more than one statement is contained in a question, it may happen that the respondent is in line only with one of them. How should the answer be interpreted in this case?
- *Create an exam-like atmosphere.* If people think that they have to take part in some sort of assessment test, they are often unwilling to respond or, even worse, choose answers that they expect the interviewer wants them to choose. This can significantly reduce the quality of the results.
- *Are too complicated.* Especially when directed to consumers, questions should not use two negations in one sentence, use much more than twenty words, contain technical terms, foreign language, etc.

Pretests are indispensable for checking the survey. They usually involve giving the questions to a small sample of respondents who are available for interview after the completion of the questionnaire. They should be asked for their impresions in order to confirm (or refute!) that the questions are clear-cut, easy to understand, and that the choice of answers accurately capture their opinions.

Consumer surveys are a powerful tool but have to be used with great care. Following the above-mentioned steps is crucial for obtaining valid results; moreover, reflecting the do's and don'ts in survey design is also helpful when evaluating the credibility of other studies.

3.3   Research in counterfeit demand

Research addressing awareness, purchase intentions, demographic characteristics or the attitudes of counterfeit consumers makes up the largest portion of the academic publications on counterfeit trade. In contrast to many studies authored or commissioned by industry associations, academic publications mostly use established constructs in their survey design (for example taken from consumer studies with respect to genuine products or from investigations into other criminal activities or socially unacceptable behavior) and refrain from using suggestive questions. In fact, many demand-side investigations provide insightful and highly relevant information for marketing and brand-protection experts. We summarize those which received considerable attention below and provide an exhaustive overview in the subsequent table.

Grossman and Shapiro (1998a) research demand-price relationships in markets with counterfeit and genuine products. However, a closer investigation of the demand-price curves reveals that their characteristic progression results from assumptions which do not reflect the critical characteristics of counterfeit products, rather, they result from depreciation effects which are not necessarily related to the phenomenon under study. In a later, frequently cited work the authors characterize counterfeiting as a disaggregation of product and brand, which can be regarded as a major contribution to theory in this field (Grossman and Shapiro 1998b).

Gentry et al. (2006) investigate product counterfeiting from a consumer search perspective. Interviews with consumers provide a picture of the cues they use to detect counterfeits and for decisions to purchase or not to purchase fake goods. Reliable cues with respect to authenticity were arrangement and location of the sales outlet and product price. While a small group of consumers said they were able to notice very minor deviations among genuine and original products, the quality only serves as an indicator for the class of poorly manufactured knockoffs. Factors positively affecting purchase decisions of counterfeits were their low prices, the low investment risks when buying low-cost fakes, negative reactions to the speed at which fashion products fall out of favor, and, in western markets, the fun of showing imitation products to friends. In China especially, the potential loss of face when exposed as a counterfeit consumer negatively affects the decision to purchase counterfeit goods.

Penz and Stöttonger (2005) rely on the Theory of Planned Behavior to systematize past findings in the field of non-deceptive counterfeit purchases and come up with a model explaining key drivers for the demand for counterfeits. Their findings indicate that search costs and accessibility are major factors determining counterfeit purchases. The embarrassment potential appeared to strongly influence purchase intentions with respect to counterfeits that are significantly cheaper than the originals, while the subjective norm (i.e. the valuation of quality and functionality) strongly influence purchase intentions with respect to counterfeits that were similar in price compared to their genuine counterparts.

Several other contributions have also concentrated on investigating customer attitudes towards counterfeits. Moores and Chang (2006), for example, find that infringements of intellectual property rights in the context of pirated software appear not to affect the perceived morality. Eisend and Schuchert-Güler (2006) thoroughly review selected studies on the determinants of consumers' intention to purchase counterfeit products and provide a theoretical concept in order to explain the motives when purchasing such goods. Furnham and Valgeirsson (2007) use Richin's materialism scale, the Schwartz value inventory, and question beliefs about counterfeits to analyze the variability in people's willingness to buy counterfeit products. They find that beliefs about counterfeits, demographic information, and materialism, but not the Schwartz value inventory (i.e. power, achievement, hedonism, self-direction, universalism, benevolence, conformity, tradition, and security), add considerably to the explanation.

Wee et al. (1995) examine the impact of non-price determinants on consumers' intentions to purchase counterfeit products. In general the authors find these determinants to be highly product specific, and thus are not able to generalize their findings. Bloch et al. (1993), in a laboratory market, offered consumers original and counterfeit goods and found pronounced intentions to purchase the illicit products. Cordell et al. (1996) also investigate the consumers' willingness to purchase counterfeit products, which they find positively related to product performance expectations. Branding and price conditions as well as retailer conditions influence the willingness to purchase low, but not high, investment-at-risk products.

Table 3.4: Demand-side investigations

Author(s)	Year	Short description
Bamossy/ Scammon	1985	– Analysis of consumer awareness, expectations and experiences of consumers who unknowingly purchased counterfeit goods. – Results from 38 interviewees reflect the limited experiences consumers had with counterfeit goods in the early 1980s.
Higgins/ Rubin	1986	– Description of the separation of brand and product for counterfeit cases where consumers purchase illicit goods knowingly.
Bloch et al.	1993	– Study on the willingness to buy counterfeits among 200 participants. – Self-image is found to be partially significant, indicating that counterfeit consumers are less confident, successful, of lower status, and less wealthy.

(Continued)

Author(s)	Year	Short description
Parthasarathy/ Mittelstaedt	1995	– Survey among 205 U.S. students. – The willingness to engage in piracy to be strongly affected by the attitude towards piracy, subjective norms, perceived utility of the software, and the willingness to seek help from others to reduce non-monetary costs. – The opinion that the high prices of software are not justified did not appear to affect piracy behavior.
Wee et al.	1995	– Empirical study among 949 Asian students. – Product-attribute variables are found to perform better in explaining purchase intentions than demographic constructs.
Chakraborty et al.	1996	– Study involving 130 U.S. students – High ethnocentrism increases the 'feeling of guilt' of consumers if the original good is produced in the U.S.
Cordell et al.	1996	– Study among 221 students, testing the correlation between: (1) the willingness to purchase counterfeit articles and the consumer's attitude toward legality, (2) the expected performance of fakes with the intention to purchase fakes in the future, (3) the dependence of risk associated with counterfeit goods and purchase intentions, (4) the consumer's likelihood of knowingly purchasing a counterfeit product, and (5) price concessions for counterfeits.
Glass/ Wood	1996	– Application of social exchange theory to investigate the influence of situational factors on the intentions to engage in software piracy. – Findings are relevant in the context of exchange in peer-to-peer networks but do not directly apply to commercial counterfeiting.
Chakraborty et al.	1997	– Examination of means of dissuading consumers from knowingly buying counterfeit articles among a group of 87 U.S. students. – Stressing inferior product quality of fakes as well as their harmful effects on national enterprises and the job market can reduce the demand for counterfeits.
Chang	1998	– Comparison of the validity of the theory of reasoned action (TRA) and the theory of planned behavior (TPB) as applied to illegal copying of software. – The results suggest that perceived behavioral control is a better predictor of behavioral intention than attitude.
Grossman/ Shapiro	1988a	– Demand-price curve in markets with both deceptive counterfeit articles and genuine products. – Welfare analysis regarding the disposition of confiscated counterfeit goods.
Grossman/ Shapiro	1988b	– Description of non-deceptive counterfeiting as a disaggregation of brand and product. – Demand-price curves in a market with counterfeit and genuine products.

(Continued)

Author(s)	Year	Short description
Tom et al.	1998	– Survey involving 126 U.S. consumers. – Results suggest that young and low-income consumers are more likely to purchase counterfeit goods than average and that satisfaction with counterfeit products is positively related to future purchase intentions.
Albers-Miller	1999	– Survey involving 92 U.S. students. – Product type, buying situation and price are found to be significant predictors of willingness to buy; the interactions of risk with respect to product type and price are also shown to be significant predictors.
Schlegelmilch/ Stöttinger	1999	– Survey among 230 U.S. students – Price difference between counterfeit and genuine products is negatively related to the intention to purchase illicit goods; quality perception of such goods and anti-counterfeiting campaigns had no significant influence on the willingness to purchase counterfeits.
Husted	2000	– The author examines the influence of national culture on software piracy based on data on 39 countries and finds piracy behavior significantly correlated with Gross National Product per capita, income inequality, and individualism. The correlation with power distance, masculinity, and uncertainty avoidance is not found to be at a significant level.
Ang et al.	2001	– Survey among 3251 Singaporean consumers with a focus on counterfeit music CDs. – Typical counterfeit consumers are value conscious, less normative, male, and of low-income.
Gentry et al.	2001	– The authors investigate choices of 102 international students between a genuine article and various counterfeits. They identify interesting aspects regarding the separation of brand and product, but do not provide an underlying analysis.
Leisen/ Nill	2001	– The survey among 144 U.S. students showed the shopping environment and the perceived financial and performance risk had a strong influence on the willingness to purchase counterfeit pain relievers, sunglasses, and watches.
Phau et al.	2001	– Survey among 100 consumers in Hong Kong. – Those who less often purchase counterfeit clothing are younger, have a lower disposable income, and are less well educated.
Swee et al.	2001	– Survey among 3,600 Asian consumers who buy counterfeit goods. – Counterfeit consumers regard the purchase of fakes as less risky and less unethical, are more value conscious, and have a lower average income compared to those who do not purchase counterfeit articles.

(Continued)

Author(s)	Year	Short description
Wagner/ Sanders	2001	– Investigation of the relationship between religion and ethical decision making for general unethical situations and purchase of pirated software. – The study suggests that religion influences counterfeit purchase decisions.
Prendergast et al.	2002	– Survey in Hong Kong of 200 consumers who have previously bought counterfeit goods. – Important factors of influence for the purchasing decision were price, the perceived level of quality, opinion of friends and family, age, money previously spent on counterfeits, and ethical and legal issues – Methods to identify counterfeits were the low price followed by the location of purchase.
Tan	2002	– Survey among 377 Chinese software users. – Purchase intentions are influenced by the perceived moral intensity, magnitude of consequences, temporal immediacy and social risk, the perceived financial, performance, prosecution and social risk, moral judgment, and cognitive moral development.
Chuchinprakarn	2003	– Investigation of the frequency of use of counterfeit goods among 662 students in Thailand. – Factors that positively affected counterfeit consumption were wealth, materialistic values and influence of celebrities.
Harvey/ Walls	2003	– Empirical study among 120 American and Asian consumers. – Price elasticities are found to be substantially larger in Las Vegas than in Hong Kong.
Hoe et al.	2003	– Study among 20 consumers under the age of 30 in the UK. – Fashion counterfeits are seen as substitutes for upscale designer brands. – Counterfeits help their buyers to create identities, communicate values and impress others. – Consumers take care that the counterfeit nature remains unknown to others.
Hung	2003	– Discussion of the roots of counterfeit trade in China. – Reasons are the strong domestic demand for imitation products and the patronage of the government.
Peace et al.	2003	– Extension of the theory of reasoned action by a factor of perceived behavioral control as posited by the theory of reasoned action, and punishment certainty / severity. – In a survey among 203 students the model was able to explain 65% of variance in software piracy intention.
Balkin et al.	2004	– The authors contend that in some cases piracy can improve the value of intellectual property and attribute this effect mainly on network and bandwagon effects and barriers to entry of competitors. However, the magnitude of these effects in comparison with negative implications for brand owners is not analyzed.

(Continued)

Author(s)	Year	Short description
Ben-Shahar/ Assaf	2004	– Development of a formal model in which a manufacturer may promote copyright infringements to indirectly participate in predatory pricing and to deter competitors from entering the market.
Moores/ Dhaliwal	2004	– Survey among 462 computer users in Singapore. The authors find that the high availability, low risk of prosecution, and high cost of the genuine software are major motives to purchase pirated copies.
Papadoulos	2004	– Investigation of the relationship between product pricing, copyright enforcement and black market formation. – Piracy is directly related to the legitimate sales price and the size of the counterfeit market.
Chiou et al.	2005	– Survey among 207 young Taiwanese consumers. – Singer idolization, perceived prosecution risk, perceived social consensus and reduced proximity reduce the attitude towards pirated music CDs.
Jenner/ Artun	2005	– Investigation of purchase intentions of 203 German tourists in Turkey with respect to textiles, fashion accessories, perfume, CDs and watches. – Expected quality of individual goods is identified as a major factor.
Penz/ Stöttinger	2005	– Survey among 1,040 Austrian consumers. – Search costs and accessibility are major factors determining counterfeit purchases. – The embarrassment potential strongly influenced purchase intentions for counterfeits which are significantly cheaper than the originals, while the subjective norm strongly influenced purchase intentions of high-priced fakes.
Wang et al.	2005	– Survey involving 314 Chinese students. – Impact of personal / social factors and attitude measures on counterfeit purchase intentions.
Cheung/ Prendergast	2006	– Investigation of the correlation between demographics of 1,152 adult Asians with their counterfeit buying behavior. – Middle and high income families, males, white collar workers, people with a high school degree and singles are more likely to be heavy buyers of counterfeit CDs. – Females are more likely to be heavy buyers of counterfeit clothing.
Eisend/ Schuchert-Güler	2006	– Review of selected studies on the determinants of consumers' intention to purchase counterfeit products.
Gentry et al.	2006	– Investigation of product counterfeiting from a consumer search perspective. – As the quality of counterfeits improves, it is becoming more difficult for the consumer to identify them.
Lau	2006	– Survey among 84 Chinese Internet users. – Extremely high prices of original software are to be seen as the main factor for the strong demand in pirated copies.

(Continued)

Author(s)	Year	Short description
Moores/ Chang	2006	– Model of ethical decision making based on the four-component model of morality. – Survey among 243 students in Hong Kong. – Use is determined by buying, buying is determined by intention, which is in turn determined by judgment. – The recognized infringements of intellectual property rights appear not to affect the perceived morality of the judgment.
Santos/ Ribeiro	2006	– Examination of the impact of Hofstede's cultural dimensions on the demand for counterfeit goods within European countries – Uncertainty avoidance and individualism are negatively correlated with counterfeiting.
Woolley/ Einingen	2006	– Analysis of purchasing frequencies and underlying antecedents of software piracy among U.S. students. – Results indicate that students' understanding and knowledge of copyright laws have increased after 1991, but this knowledge has not influenced software piracy rates.
Bian/ Veloutsou	2007	– Comparison of British and Chinese consumers who admit to have knowingly purchased counterfeit products. – Demographic variables have not been found to significantly influence counterfeit demand. – In both countries consumers show a very low opinion of counterfeit products in general, but perceive their average quality as similar compared to non-logo products.
Furnham/ Valgeirsson	2007	– Study on the foundation of Richin's materialism scale, the Schwartz value inventory, and questions concerning beliefs about counterfeits. – Beliefs about counterfeits, demographic information, and materialism, but not the Schwartz value inventory, add considerably to the explanation of the variability in people's willingness to buy counterfeit products.
De Matos et al.	2007	– Survey among 400 Brazilian consumers. – Integration of the main predictors of the consumers' willingness to purchase counterfeit goods in one structural equation model.

The difference in purchasing behavior of consumers in test markets with and without imitation products has not yet been thoroughly analyzed. Corresponding findings would help to understand the interference of counterfeit trade with licit trade and thus allow for a better understanding of substitution effects with respect to non-deceptive counterfeiting. Another area of study which has only been vaguely investigated is the impact of illicit markets on the perception of trademarks and brands and thus on the goodwill towards licit enterprises. Both areas need to be addressed when conducting financial impact analyses.

Part B   Countermeasures – Best Practices and Strategy Development

4   Established Anti-Counterfeiting Approaches – Best Practices

Established brand- and product-protection strategies do not seem to sufficiently reflect the complexity of the counterfeit market. Governments and companies have not been able to stop the growth in counterfeiting, and the elaborate production and shipment tactics of illicit actors challenge even the most experienced anti-counterfeiting experts.

The commitment to tackle the problem is without doubt there. Companies are investing considerable effort to try and implement adequate brand- and product-protection strategies. However, their market insights as well as their know-how on organizational and technical approaches are mostly limited to their experience from within their own company and, at best, to some informal exchange among colleagues, for example at brand-protection conferences. Strategy development would greatly benefit from knowledge transfer within industries and among different industry sectors.

The following chapter tries to close this knowledge gap. We detail selected findings from a recent cross-industry benchmarking study in which we identified successful strategies of leading brand owners and manufacturers. Following this study we outline the characteristics of successful monitoring, prevention and reaction measures.

4.1   State-of-the-art in anti-counterfeiting

Benchmarking studies serve as tools in which organizations evaluate various aspects of their processes in relation to best practice. In general such analyses can be conducted among departments or business units within a single company, among enterprises within an industry sector, or, as in our case, among organizations from different industries. A comparison of the participants helps to identify strengths and weaknesses of the process under study and thus to develop plans on how to improve their performance. The studies foster learning effects and enable companies to overcome "paradigm blindness" by exposing them to other working solutions. Beyond the learning effects among companies, benchmarking studies allow researchers to feed in information from related areas (for example with respect to gray markets, drug trafficking, theft, but also general aspects of process design, innovation management, etc.) and codify their insights into current business practices.

The properties of actual anti-counterfeiting measures were investigated within a benchmarking study among 45 carefully selected companies (out of a group of approx. 210 leading enterprises with a global market presence). The study was conducted between April 2006 and January 2007. In order to structure the investigation, the object of research was divided into four areas: existing knowledge, prevention measures, reactions, and monitoring processes (c.f. Figure 4.1).

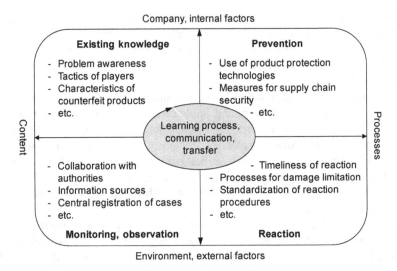

Figure 4.1: The structure of the company survey

Companies have been studied with respect to all four categories. After the first pre-selection a questionnaire-based survey was used to identify those manufacturers and brand owners that have implemented cutting-edge anti-counterfeiting measures in at least one of the above-mentioned categories. The questionnaire-based survey was followed by semi-structured interview rounds to better describe best practices from different industries.

Questionnaire design

The empirical data stems from a questionnaire-based survey. The approach was preferred over personal interviews to give respondents time to provide thoughtful answers and to allow them to consult with others or look up records. The constructs were designed together with industry experts after a thorough literature review. Overall, 60 items were tested based on a five-point Likert scale. The study design facilitated the calculation of individual scores during the evaluation process. Six and seven additional constructs allowed for multiple selections and numerical answers respectively, for example to identify departments, geographic regions, or the number of employees.

Pretests helped to evaluate the comprehensibility of the questions and to measure the time needed to answer them. On average the completion of one questionnaire took 30 minutes, which was regarded as a good compromise between the level of detail and required effort on the side of the respondent. The questionnaires were made available in both English and German language and could be completed either electronically or on paper.

The survey was addressed to the people in charge of anti-counterfeiting-related activities; the individuals had been identified in preceding telephone calls conducted by an experienced benchmarking team. Each potential respondent was contacted in advance to increase the response rate.

Sample selection

Care was taken to pre-select those companies with extensive experience in the fight against counterfeit trade. Therefore the sample consists of the 50 (mostly multinational) companies whose brands head the list of the most frequently seized counterfeit articles provided by European customs. In addition to this group 143 companies listed in the German stock exchange indices DAX 30, MDAX, TECHDAX, and the Swiss index SMI were chosen if they (1) produced or marketed products similar to the above-mentioned frequently targeted goods, or (2) produced or marketed products which are counterfeited less often, but constitute a severe risk for the consumer or user when sold as fakes (for example counterfeit pharmaceutical products or aviation spare parts). 12 privately-held companies whose characteristics were similar to the above-mentioned group members were addressed as well.

Overall, 22% (45) of the questionnaires were completed and returned, which is a high rate of return for a mostly confidential and otherwise undisclosed topic. Figure 4.2 and Figure 4.3 summarize the characteristics of the respondents.

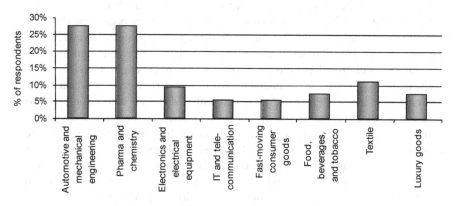

Figure 4.2: The characteristics of the respondents by industry type

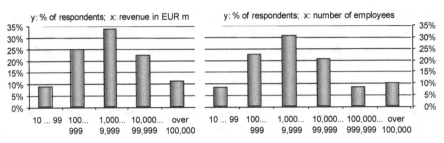

Figure 4.3: The characteristics of the respondents by industry type

Evaluation criteria

Among the participants a ranking was compiled based on the cumulated score of the self-reported (and, if possible, scrutinized) capabilities in each category. In the following the characteristics of the eight "successful practice" companies are compared against the characteristics of the rest of the group[19] in order to stress meaningful differences, as well as to outline the potential success factors.

Evaluation and findings

Of the respondents, 71% reported that for their organization dealing with the problems of product and trademark counterfeiting has significantly gained in importance over the last five years. Depending on the industry, companies were either particularly afraid of endangering the consumer's health and safety or of losing revenue (c.f. Figure 4.4). The expected impact on the brand as a whole (and thus on brand value) appeared to be a common denominator across all industries.

At first sight the differences of the responses of the top 8 companies and the rest of the group seem to be only marginal. However, the top 8 companies were less likely to take intermediate positions when evaluating the impact on revenue and consumer safety, which seems to reflect a thorough understanding of the implications.

The ability to quantify the financial impact of counterfeit trade, however, seems to be rather limited (c.f. Figure 4.5). Overall, only 19% of the respondents claimed to possess good estimates on the loss of revenue due to counterfeit trade (two of them among the top 8 companies), and 5% reported they had reliable estimates for the loss of brand value (one of the top 8 companies). Telephone interviews following the survey revealed that, in fact, none of the manufacturers and brand owners were able to provide substantiated quantitative analyses that could help to calculate a business case for example for anti-counterfeiting technologies.

---

[19] Please note that, due to the pre-selection process, non-top-performing companies also have above-average anti-counterfeiting measures in place.

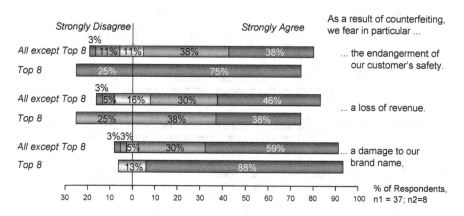

Figure 4.4: The perceived implications of counterfeit trade

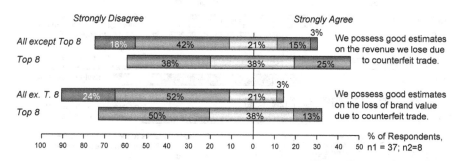

Figure 4.5: Ability to quantify the impact of counterfeit trade

Unlike the restricted ability to assess the impact of counterfeit activities at a quantitative level, most of the participating companies possess good tactical knowledge of the counterfeit market, including knowledge of country-specific characteristics of counterfeit trade, import routes, and the typical level of quality of the imitation products (c.f. Figure 4.6). In each category the top 8 companies performed better than the other respondents.

With respect to well-defined[20] anti-counterfeiting-related processes, the group of the top 8 companies differed considerably from the remaining respondents (c.f. Figure 4.7). All top 8 companies have developed and implemented defined processes which govern their reaction to the emergence of counterfeit articles, whereas

---

[20] With well-defined processes we mean processes that are described in written form, such that employees can follow guidelines or are supported by checklists in case of counterfeit occurrences.

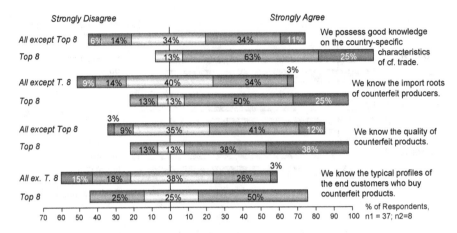

Figure 4.6: Knowledge of the supply-side of counterfeit trade

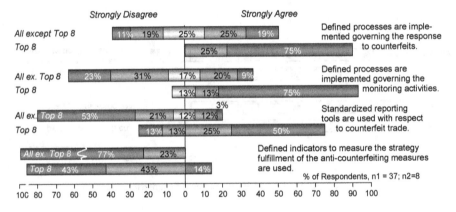

Figure 4.7: Standardized processes, monitoring, and reporting tools

only 44% of the other respondents had such processes in place. Similar results were found regarding the companies' monitoring activities. In general the existence of defined processes allows brand owners and manufacturers to more easily extend anti-counterfeiting measures, for example by introducing them to other business units or by adding people to the process without the need to let them develop detailed tactical knowledge or experience.

Standardized reporting tools are an important means to obtain reliable market data and thus allow for an evaluation of the development over time and, accordingly, a comparison between different analyses. Moreover, the existence of standardized reporting schemes seems to correlate with a high commitment of senior management. While six of the top 8 companies made use of standardized reporting, only 15% within the other group had such tools in place.

Hardly any of the companies seemed to possess or use defined indicators to measure the strategy fulfillment of the anti-counterfeiting measures. This finding underlines that well-defined, process-oriented anti-counterfeiting efforts are still part of a very young field of activity for most enterprises, with the strategies and goals still being rather vaguely formulated.

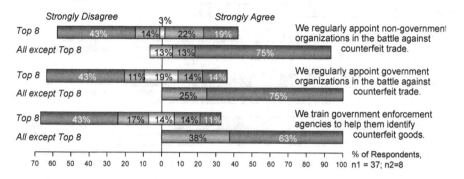

Figure 4.8: Cooperation with external stakeholders

Figure 4.8 shows the share of companies which regularly appointed external (i.e. government and non-government) agencies to support their product and brand-protection activities. While the integration of external stakeholders was a common practice among the top 8 companies, it was an exception rather than the rule among the others. The same was true for training government enforcement agencies in detecting and identifying counterfeit articles. Telephone interviews with the top 8 companies showed that these enterprises have regularly provided customs officers with counterfeit-related information and offered training sessions either directly or via industry associations.

Figure 4.9 provides an overview of the perceived importance of individual anti-counterfeiting measures. The participants rated protection technologies, legal actions, organizational supply chain security measures, organizational measures to secure the distribution system, and activities to increase problem awareness among both consumers and political decision makers.

Protection technologies (such as holograms, micro-printings etc.) as well as legal actions were attributed only below-average performance. However, the top 8 companies were slightly more optimistic with respect to technological approaches than the rest of the group, but less confident on legal measures. Interviews with the respondents showed that the latter are primarily seen as a pre-condition to take further steps when counterfeit cases occur but are as such not effective at fending off counterfeit trade.

Organizational steps to enhance supply chain security or to secure the distribution system were rated as very important by most respondents. The top 8 companies attached even more importance to such measures that include, for example,

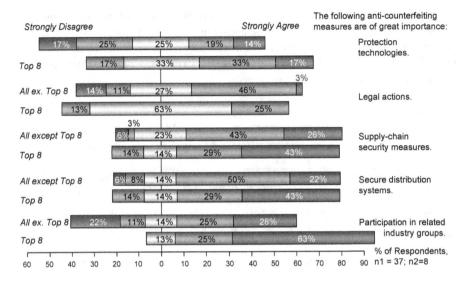

Figure 4.9: Key measures to fend off counterfeit trade

evaluation of suppliers, awareness raising campaigns at distributors, or the choice of specific distribution channels for high-risk products.

The most pronounced difference appeared with respect to activities that increase awareness and help to gain government support. While participation in related industry groups was regarded as very important by 63% of the top 8 companies, only 28% of the other group said so. Influencing demand and lobbying for a more effective enforcement of intellectual property rights also in Asia and Eastern Europe is in fact promising as it deals with the problem at its roots.

The reported influence of counterfeit trade on brand positioning, selection of the sales markets, and outsourcing activities is illustrated in Figure 4.10. The findings are surprising and appeasing at the same time. None of the top 8 companies claimed that the counterfeit trade had an impact on related strategic decisions, while 26% of the other companies acknowledged the existence of such an influence. However, the finding is only ostensibly counterintuitive. Interviews among the top 8 companies revealed that these organizations regard counterfeiting as one important issue among others, rather than as a disruptive phenomenon. That does not mean that the existence of imitation products will not influence decision making – for example on how a product is marketed or which sales channels are chosen – but puts the problem in the realm of other issues which, as further interviews revealed, are *just* to be taken care of. A successful anti-counterfeiting strategy prevents a company from having to take more drastic countermeasures, such as the withdrawal from certain markets or the integration of upstream or downstream production processes.

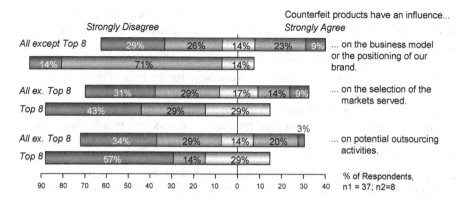

Figure 4.10: The influence of counterfeit trade on strategic decisions

4.2   *The characteristics of successful practices*

Successful anti-counterfeiting strategies are highly brand- and product-specific; they have to reflect the tactics of counterfeit actors, consumer demand and awareness, the major functions of the brand, the role of enforcement agencies in the country of production and sale, the potential consequences of counterfeit occurrences, etc. However, a number of successful practices appear to recur regardless of the industry or the type of product. We briefly highlight each aspect below and discuss how to put the resulting recommendations into practice in the Chapter 5.

Knowledge and market insights

The ability to successfully deal with counterfeit cases correlates strongly with the presence of above average market insights. All top performing companies that we worked together with had in-depth knowledge on the country-specific characteristics of counterfeit trade and the production techniques of illicit producers, a precise idea of the different tactics to disguise criminal activities and conceal import routes, and the typical distribution channels and selling tactics. Moreover, the insights were not concentrated in only one or two people but, at least to a certain extent, available in written form. Companies had reports, guidelines, and background information at hand that they made available to selected employees. By appointing anti-counterfeiting experts in different business units and at different locations, companies established an information exchange from and to their various experts. This form of knowledge transfer supports a strategy of local monitoring activities as well as reaction measures that are strengthened by established contacts to local, external stakeholders (for example to customs officers, wholesalers, etc.). At the same time it allows for having a complete picture on the development of the counterfeit market, coordinating activities within the company, for example to initiate a public relations campaign following a counterfeit incident, and to allocate funds to brand-protection activities.

Other common building blocks were workshops for the staff of the companies' purchasing and selling departments as well as information sessions for component suppliers and foreign distributors. While many brand owners still avoid brand-specific campaigns to raise consumers' problem awareness, best practice companies more often use their websites to provide information on how to identify counterfeit articles or spot risky sales channels.[21]

It goes without saying that knowledge as such does not directly translate into best practices, but we want to stress its importance for their development. Simply copying the working solutions of others is not enough to solve such a complex problem in the long run. When defining the general terms of reference for a brand- and product-protection task force, one should keep in mind that the strategy has to evolve alongside the development of the counterfeit market. Therefore the employees have to be sensitive to such changes. Effective transfer of knowledge among different functions within the company as well as careful integration of external stakeholders must be a clear objective of the strategy development process.

Process design

In many companies we worked together with, the anti-counterfeiting initiative developed along the following lines: (1) the problem was identified, often in consequence of a counterfeit case affecting the company directly or one of its competitors; (2) an employee (mostly from the legal or security department) was assigned to deal with similar problems in the future; (3) a growing number of counterfeit cases (often due to an absolute increase and better monitoring activities) led to extended brand- and product-protection activities; (4) the problems became more visible within the company (due to awareness training among employees, management reports, etc.); (5) a larger number of employees in different locations and business units became involved and finally, (6) the decision was made to structure and formalize the anti-counterfeiting activities. The elaboration of well-defined processes that govern reaction and monitoring activities as well as the deployment of standardized reporting tools may be seen as a logical consequence of this development. However, we want to point out their importance for a successful coordination of anti-counterfeiting measures. Meaningful trend analyses only become possible with standardized monitoring and reporting tools, for example in the form of predefined reporting sheets with clear guidelines how and

---

[21] Publishing brand-specific information on counterfeit trade can not only prevent accidents but may also help to fend off liability claims in case of physical injuries following the consumption or usage of substandard imitation products.

when to estimate the market share of counterfeit goods. With step-by-step guidelines on how to respond to counterfeit occurrences, employees who do not encounter illicit products on a day-to-day basis can at least initiate the first steps. Best practice companies tend to have emergency plans at hand to respond quickly and routinely to the occurrence of counterfeit articles, including frameworks for press releases, contact lists for product recalls, etc. In fact, successful anti-counterfeiting measures are guided by well-defined process steps. They should again encourage local action and support central control. Such processes require cost sharing if, for example, anti-counterfeiting measures benefit the entire company but have to be initiated by individual branches. Most companies that scored well in the benchmarking study paid for most anti-counterfeiting measures that are related to monitoring and reaction (especially legal expenses) from the budget of a central brand-protection task force but made the individual business units pay for (mandatory) security features.

Some common elements of successful anti-counterfeiting strategies

Successful anti-counterfeiting strategies are composed of a set of different, partly complementary and partly overlapping practices. No single "silver bullet" seems to exist. Companies often rely on a combination of organizational, technical and legal measures together with campaigns to inform various stakeholders. While legal steps (i.e. the registration of specific designs and the use of authentication technologies) are mostly seen as a basic prerequisite, companies set priorities to organizational measures that ensure the integrity of their parts supply and strengthen the security of their distribution system. Authentication technologies are used to support the organizational steps. Very high hopes are placed on both lobbying for better protection of intellectual property rights and consumer information to raise problem awareness and reduce demand for illicit goods. Both are done primarily with the help of industry associations to avoid negative associations with a brand. Moreover, successful companies routinely appoint non-government organizations such as private detectives to support their investigations and also involve government organizations, in particular customs, in their activities.

4.3   Research on managerial and legal countermeasures

Managerial countermeasures have received some attention among researchers, which reflects industry's demand for effective anti-counterfeiting strategies. Harvey and Ronkainen (1985) identify some rudimentary strategies that companies use to combat counterfeit trade: hand-off, prosecuting, withdrawal, and warning. Shultz and Saporito (1996) provide a more detailed framework of anti-counterfeiting strategies, including measures such as "use of high-tech labeling", "co-opt offenders", and "educate stakeholders at the source", but do not suggest how to operationalize these strategies.

Chaudhry et al. (2005) address the questions of how managers can conceptualize the intellectual property environment, how such environments affect market-entry decisions, what anti-counterfeiting strategies are frequently used and how efficient each tactic is in the host country market. Bach (2004) examines the interaction of law and technology – the "simultaneous and interwoven deployment of legal and electronic measures to protect digital content" – and raises the question of whether technologies are merely a defense strategy against piracy, as the industry asserts, or rather an attempt to fundamentally redefine the producer-consumer relationship.

Hung (2003) finds that not many companies are able to oppose counterfeit trade on a managerial and legal basis. He supports this with the example of the Chinese environment where the response strategies, which are recommended in business literature, only show limited effect. He concludes that foreign companies may have to wait until China becomes a victim, on balance, instead of a benefactor of product counterfeiting before they can rely on better protection of their intellectual property.

Olsen and Granzin (1992) investigate licit supply chain members other than brand owners regarding their actions towards illicit trade. Their study is based on data gathered from store managers to examine responsibility, willingness to help, and possible conflicts with respect to counterfeit products. The findings are of importance when assessing the acceptance of anti-counterfeiting strategies.

Without doubt, the existing literature aims to provide guidance for practitioners who have to define anti-counterfeiting strategies. Recommendations, however, are mainly based on observations of established practice and are only rarely directly derived from the characteristics of the counterfeit market. An understanding of the motives, production settings, strengths and weaknesses of the illicit actors would benefit the licit stakeholder when defining anti-counterfeiting measures. Moreover, performance measures for anti-counterfeiting activities could help to identify, select and improve successful practices.

Table 4.1: Academic literature on the management of anti-counterfeiting activities

Author(s)	Year	Short description
Kaikati/ LaGarce	1980	– Discussion of different forms of brand piracy. – Strategies to prevent counterfeiting. – Outline of international laws to protect trademarks.
Harvey/ Ronkainen	1985	– Discussion of potential ways illicit actors can obtain classified information which enables them to produce counterfeit articles. – Loss estimates based on industry estimates.
Harvey	1987	– Potential ways illicit actors can obtain confidential information – Potential corporate responses to counterfeit occurrences.
Harvey	1988	– Discussion of an organizational structure for a 'Counterfeit Prevention Task Force' involving employees from marketing, research, and development.

(Continued)

Author(s)	Year	Short description
Bush et al.	1989	– Survey among 103 companies. – Discussion of the implications of the 1984 Trademark Counterfeiting Act and company-internal actions.
Olsen/ Granzin	1992	– Depiction of how manufacturers can establish a relationship with their distributors to gain support in fighting illicit trade. – Interviews with five retailers from the automotive industry to conceptualize a structural equation model.
Olsen/ Granzin	1993	– Investigation of the influence of dependence, control, channel conflict and satisfaction concerning a dealer's willingness to help a manufacturer combat counterfeiting. – Findings are that manufacturers can engender cooperativeness by nurturing satisfaction and dependence in manufacturer dealer relationships.
Chaudhry/ Walsh	1996	– Overview of the legal framework to avert counterfeit trade, reviews anti-counterfeiting strategies (termed warning, withdrawal, prosecution, awareness, assertion).
Nill/ Shultz	1996	– Discussion of ethical decision making, moral reasoning and purchase intentions in the context of counterfeit goods.
Shultz/ Saporito	1996	– Discussion of strategies to respond to counterfeit producers.
Simone	1999	– Discussion of the difficulties for brand owners to pursue remedies against infringers in China.
Krechevsky	2000	– Proposal of practical anti-counterfeiting measures for affected enterprises.
Jacobs et al.	2001	– Discussion of strategies to deal with counterfeit trade.
Green/ Smith	2002	– Summary on the literature that addresses counterfeit trade. – Strategies for addressing the threat in developing markets. – Case study of a major company producing and selling alcoholic beverages.
Trainer	2002	– Discussion of the hurdles of brand owners who, despite an impressive number of IP laws in China, often face reluctant prosecutors and ill-trained judiciary.
Hung	2003	– Discussion of the roots of counterfeit trade in China. – Reasons are the strong domestic demand for imitation products and the patronage of the government.
Yang et al.	2004	– Elaboration of ten corporate actions to avert counterfeit trade.
Bach	2004	– Discussion of the implications of music piracy of consumers and the control of media companies through electronic monitoring. – The point of view is expressed that both extreme scenarios (total devaluation of intellectual property vs. total control of the right holders) are unlikely and that a careful discourse of policy makers is necessary in order to find a suitable intermediate position.
Javorcik	2004	– Investigation of the relationship of the degree of enforcement of IPR on the composition of foreign direct investment. – A lack of IPR protection deters investors from undertaking local production but encourages them to focus on distribution of imported products.

(Continued)

Author(s)	Year	Short description
Chaudhry et al.	2005	– Investigation of how managers conceptualize the intellectual property environment; how the intellectual property rights environment affects the market-entry decision; what anti-counterfeiting strategies are frequently used; and how effective each tactic is in the host country market.
Sonmez/ Yang	2005	– Single case study based on Manchester United Football Club's efforts to establish its trademark licensing and to deal with counterfeit products in China.
Chaudhry	2006	– Summary of statistics on product categories and countries of origin of counterfeit goods as well as surveys on the attitude of consumers towards counterfeit articles – Overview of current enforcement initiatives in the EU and the U.S.
Clark	2006	– Outline of selected tactics of counterfeit producers to disguise their operations, changes companies should make, and issues the Chinese government should address.
Wald/ Holleran	2007	– Discussion of a gray market and diversion problem at Johnson & Johnson's Medical Device & Diagnostics in 2001.

Legal issues

Intellectual property rights (IPR) are key issues for international trade. Consequently, the academic literature on IPR is very comprehensive. Rather than aiming to provide an exhaustive overview, we refer to Maskus (2000) for a thorough introduction, and concentrate on selected publications which bear a direct reference to counterfeit trade.

Globerman (1988) briefly discusses the cost of trade protection. He recommends a policy approach that emphasizes the 'private' efforts of brand owners to protect their products rather than strengthening retaliatory trade legislation. In this context, Hetzler (2002) summarizes the legislative measures taken in the EU between 1988 and 2002. He highlights the effects of these initiatives for the German market and elaborates on the relationship of counterfeit producers with organized crime, emphasizing the entrepreneurial character of the activities.

Taking a global perspective, Jain (1996) investigates the conflict between the industrialized and developing countries which favor a high and a low level of protection, respectively. Shultz and Nill (2002) further elaborate on this issue by introducing a game theoretical perspective to examine violations of IPR within the context of social dilemmas. The contradicting interests of industrialized and less developed countries are exemplified using the prisoner's dilemma.

In his empirical study, Javorcik (2004) explores the relationship of the degree of enforcement of IPR on the composition of foreign direct investment. The data indicates that investors relying heavily on the protection of IP are deterred by weak IPR regimes. A lack of IPR protection deters investors from undertaking local production but encourages them to focus on distribution of imported products. The latter effect is present also in sectors that do not heavily rely on IPR protection.

Table 4.2: Academic literature on legal issues and legislative concerns

Author(s)	Year	Short description
Globerman	1988	– Discussion of net costs of counterfeit trade and the cost of trade protection. – Policy approach that emphasizes the 'private' effort of brand owners to protect their products rather than strengthening retaliatory trade legislation.
Jain	1996	– General issues of intellectual property infringements and the conflict between the industrialized and developing countries.
Ronkainen/ G.-Cusumano	2001	– Determinants of intellectual property violation on the example of software piracy.
Hetzler	2002	– Legislative measures taken in EU between 1988 and 2002 and their effects for the German market. – Relationship of counterfeit producers with organized crime.
Mitchell/ Kearney	2002	– Discussion of measurement techniques of brand confusion to better evaluate and judge upon trademark infringements.
Shultz/ Nill	2002	– Game theoretical perspective by examining violations of intellectual property rights within the context of social dilemma.
Javorcik	2004	– Investigation of the relationship of the degree of enforcement of IPR on the composition of foreign direct investment. – A lack of IPR protection deters investors from undertaking local production but encourages them to focus on distribution of imported products.
Liebowitz	2005	– Positive effects of counterfeit products on licit brand owners in the form of indirect appropriability, the exposure effect, and network effects.

5 Implementing Anti-Counterfeiting Measures

In Part A we outlined the characteristics of the counterfeit market with a focus on both supply and demand of illicit goods. These analyses were essential to obtain a better understanding of the different illicit business models, the corresponding production and distribution strategies and the role of the consumers. They also allowed us to identify the strengths and weaknesses of counterfeit actors and the drivers and enablers of counterfeit demand. In Chapter 4 we described current best practice with respect to brand- and product-protection from a company's perspective. Chapter 5 now combines the market insights with knowledge of state-of-the-art countermeasures. We describe how to design and implement effective monitoring, reaction, and prevention processes and outline how an adequate organizational structure of a brand- and product-protection program can look.

5.1 Monitoring processes

Monitoring of counterfeit market activities serves two purposes. Firstly it helps to identify incidents of counterfeit trade and allows for timely responses at an operational level. Secondly it provides management with information on the development of the counterfeit market and thus supports decision-making as do market observations or competitor analyses for the licit market.

Though these two principle objectives are the same for all related monitoring activities, their set-up varies considerably between different companies. The basic design parameters are (1) the average number of counterfeit cases per unit of time (i.e. the frequency of counterfeit occurrences) and (2) the risk that emanates from individual imitation products. In the case of rare but possibly high-impact incidents (for example counterfeit brake pads), monitoring activities utilize listening posts to capture information from customs, supply chain partners and consumers. Moreover, information exchange across industry associations is desirable as an illicit producer often targets more than one brand within the same product category. For rare but high-impact incidents, implementing monitoring processes is closely related to awareness training. Employees at different locations and in different functions, etc. as well as business partners outside the company have to have the background knowledge to identify "peculiar incidents" and should know whom to contact to initiate further investigations.

If a larger number of faked products are sold on a regular basis, illicit actors need to establish some sort of stable distribution channel to market their products.[22] In such cases brand owners must obtain timely data on shipment routes, sales channels and identifying features of imitation products to ease product inspections at other locations. Monitoring activities often call for the engagement of local investigators to track down middle men and get to the sources of counterfeit production. Local contacts can also support test purchases in suspect stores or help to analyze warranty claims, etc. Frequent counterfeit cases also lead to better market insights, thus helping to identify high-risk distribution channels and typical import routes. Monitoring with respect to frequently targeted product categories therefore resembles market research activities alongside formalized observation routines of previously identified high-risk spots. It requires a great deal of information sharing among the stakeholders. The relevant information sources, as well as selected recommendations for organizing data collection and reporting activities, are outlined below.

Information sources

Several data sources allow for the identification of counterfeit occurrences, the compilation of quantitative analyses on the market share of counterfeit goods and a better understanding of the mechanisms and development of the illicit market. We outline below their value for non-quantitative analyses and describe several ways to gain access to the required information. Selected sources will be taken up again later when introducing counterfeit market share analyses.

- *Customs enforcement statistics* not only provide the basis for counterfeit market share analyses. They also reveal other information such as locations of production, import routes, and addresses.[23] Enforcement statistics are instrumental in gaining an understanding of the market trends and may help to predict the future development of counterfeit activities (related to a company's own as well as the competitors' products), which makes them an integral part of most annual or semi-annual reporting activities. Even for non-European companies it is worth visiting the EU customs website for detailed seizure statistics on individual brands, product, countries, etc.[24] However, official statistics rarely provide timely information and are thus only of limited value for triggering reaction processes.
- *Consumer surveys* can provide a variety of information on market penetration and the availability of imitation products, pricing strategies of counterfeit actors,

---

[22] They may also use established, licit distribution channels but the implication with respect to monitoring processes are the same.

[23] The address will most likely not directly reveal the identity of the recipient, but the shipment has to contain some sort of information on where the products will be delivered to.

[24] See http://ec.europa.eu/taxation_customs/customs/customs_controls/index_en.htm.

sales channels, consumer behavior and awareness, reasons for and against buying counterfeit goods, characteristics of counterfeit consumers, the perceived impact on the exclusiveness and reputation, etc. Brand owners should regularly conduct consumer surveys, particularly in the case of product categories which are frequently targeted by illicit actors. However, surveys are difficult to design as they are likely to address socially unapproved behavior. They are also expensive to conduct if the required sample size has to be high. A comprehensive empirical study and its design have already been outlined in Chapter 3.

- *Sampling* connotes the collection of data from the execution of test purchases. It provides the brand owner with information on how counterfeit goods are sold (for example appearance of the store, typical customers of the store, etc.), and allows for a detailed physical analysis of the products after purchase. Characteristics of interest especially include the visual and functional quality, used materials, potential risks for the user or consumer, and the way illicit producers deal with anti-counterfeiting technologies. Moreover, test purchases help to find identifying features that facilitate the search for further faked products, allow for quick intervention at the point of sale, and signal that the brand owner takes fighting illicit trade seriously. However, the approach is relatively expensive especially if the authentication procedure is destructive[25] and if the share of counterfeit articles is small. Often goods confiscated by customs or articles which have been returned to the manufacturer as warranty cases provide a cheaper alternative to sampling. The use of counterfeits that have been collected by other means is therefore preferred if the approach does not introduce an unknown bias and if information on the purchasing environment is not of primary interest for the brand owner.

- *Company-internal data*, for example on regional sales or warranty cases, can provide timely information on counterfeit occurrences. It can therefore support decisions to trigger reaction processes. While changes in regional sales may be an indicator for (mostly non-deceptive) counterfeits that are sold in large quantities (for example tobacco products), an analysis of warranty cases or consumer inquiries constitutes a more sensitive instrument. It is, however, only applicable with respect to deceptive counterfeit cases where customers are concerned about unannounced changes in product appearance, taste, etc. The use of company-internal data presumes an information exchange with departments which were not necessarily directly concerned with imitation products in the past. Therefore campaigns to raise awareness across different departments of the company, including customer support, are a key factor for a well-functioning monitoring system.

- *Data on counterfeit production* facilities helps to gain a better understanding of the mechanisms of counterfeit trade. Related information, for example on the output of such facilities, facilitates retrospective cross-checks of other

---

[25] The articles mostly have to be unpacked for the test and can often no longer be sold thereafter. Moreover, the test itself may be time consuming and thus expensive.

monitoring activities (one may ask questions such as "how many products that have very likely been produced here have we confiscated?").

- *Information provided by supply chain partners* (including customs), customer, and consumers constitutes a major source for timely cues on counterfeit occurrences. Within the corporate world and especially with respect to organizations which are not directly affected by a given counterfeit case, established personal contacts seem to be essential for thorough information exchange. Downstream supply chain partners who may find themselves competing with counterfeit articles should be provided with contact information and be frequently encouraged to provide feedback – at least they have a vital interest in helping to combat fakes. In order to receive tip-offs from end users, entry points can be established, for example via trained personnel at consumer hotlines who may forward individual calls to dedicated employees if suspicion is raised. Interviews with brand-protection practitioners indicate that the majority of reaction processes are triggered by tip-offs provided by these data sources, which emphasizes the importance of "listening posts" in supporting anti-counterfeiting efforts.
- *Information from competitors* often proves helpful as counterfeit actors frequently target more than one product or brand. Industry associations play an important role in establishing contacts among brand-protection experts.
- *Information from Internet research* should be an inherent part of the monitoring process. The World Wide Web constitutes an important distribution channel especially for non-deceptive counterfeits (for example watches, jewelry, and fashion accessories) and for deceptive counterfeit products that consumers tend to avoid purchasing in person (for example erectile dysfunction medication). A number of automated tools exist that can be configured to automatically search for dubious websites. We describe several of them in Info Box 5.1.

In most cases there is no single information source that sufficiently supports a company's monitoring process alone. Anti-counterfeiting experts should consider using all possible ways to obtain information on counterfeit occurrences in the early stages of their efforts and may decide later which source to omit or use more extensively. Monitoring should be seen as an iterative process that has to be reviewed periodically. The process may be labor-intensive as it requires many different stakeholders at different locations and in different departments to work together. However, it also helps a great deal in the process to increase problem awareness throughout the company. Without proficient monitoring activities the entire brand- and product-protection effort is unsettled.

Info Box 5.1: Automated Internet search engines

In countries where it is too risky to display counterfeit goods in public, the Internet constitutes an important sales channel for illicit actors. There, they can set up and relocate online stores in no time, sell products over online auction sites using numerous pseudonyms, and use spam emails to draw attention to their goods. For brand owners, searching the web for infringements is only feasible using automated tools. These tools assist trademark owners in identifying sites that misuse brand names, offer diverted goods or counterfeits, falsely claim relationships, or associate the brand with objectionable content. The tools look for the respective brand names and words that are orthographically or phonetically similar and that are displayed together with keywords like "authentic", "cheap", "discount", "genuine", "factory overruns", or "gray imports". They may also flag colors that the original product is not made in, atypical sizes and quantities, and prices that are far too low. The tools need to be fed with information on the corresponding brands and may have to be trained in order to generate reliable reports. However, they facilitate analyses of huge numbers of sites each day, generate frequent reports and statistics, and may even automatically send out warnings to the infringer or initiate steps to block an illicit shop's Internet domain. Companies that offer monitoring tools for brand-protection purposes are, among others:

- CSC Corporation Service Company, www.cscprotectsbrands.com
- Cyveillance Inc, www.cyveillance.com
- Envisional Ltd, www.envisional.com
- Kessler International Ltd, www.investigation.com
- MarkMonitor Inc, www.markmonitor.com
- OpSec Security Group plc, www.opsecsecurity.com
- Partners 4 Management GmbH, www.p4m.de
- Sublime IP Ltd, www.sublimeip.com
- Trade Safeguard LLC, www.trademarksafeguard.com

For brand owners that face counterfeit trade in the consumer market, such search engines should be an indispensable part of their monitoring efforts.

Reporting activities

At a higher level of aggregation, data regarding the market share, the visual and functional quality of the counterfeits, the effect of security features, the distribution of sales prices and information concerning consumer behavior to assess the share of deceptive and non-deceptive fakes is required. The information obtained from the various sources has to be organized, processed and put into a form which is suitable for presentation and communication to a given audience that does not necessarily consist of anti-counterfeiting experts. If reporting activities relate to internal company communication, they are supposed to provide management with unbiased information to support the decision-making process. With respect to illicit market activities they should include an overview of the development of the illicit activities and a brief summary of the recent cases the company has been involved in. Demand-side-related information, for example on consumer awareness, willingness to purchase, or the perceived impact on the reputation or exclusiveness of the brand, should also be included. Figure 5.1 exemplifies a simple reporting sheet that has been developed within a recent research project.

An assessment at regular intervals and in a standardized form allows for the discovery of trends and developments which will feed back to the impact analysis and can refine the monitoring process. The frequency of reporting depends on the magnitude of the threat. Frequently-targeted enterprises should at least compile semi-annual reports to capture the dynamics of the counterfeit market. In any case

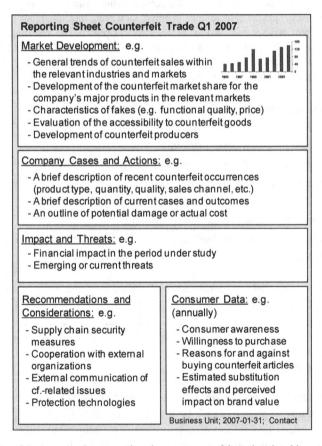

Figure 5.1: An exemplary reporting sheet on counterfeit-trade-related issues

a concisely defined set of rules should underlay all quantitative statements to ensure the comparability of the results (we introduce suitable measurement techniques in Chapter 6). Besides the importance for the decision-making process, standardized reporting tools may also help to increase awareness of the problem among senior management, provide an organ for brand-protection experts, foster information exchange between different business units and functions, and help to document the company's anti-counterfeiting efforts.

5.2 Reaction processes

Companies' reaction to counterfeit occurrences can start at ignoring individual imitation products and escalate to broad-ranging reactions involving enforcement agencies from several countries, seizures, and product recalls. Trademark and copyright holders have the prime responsibility for initiating measures to protect their goods, and management has to decide which thresholds should be applied for further action and which steps should be taken if action is necessary. The decision should be based on an analysis balancing the cost of reaction and the impact of the illicit goods potentially being sold.

The reaction process in general may be subdivided into six steps: (1) withdrawing counterfeit articles from circulation, (2) informing/warning those who are affected by the imitations, (3) finding and if possible eradicating the source of the illicit products, (4) seeking the prosecution of offenders, (5) managing the relationship with the informant and other supportive stakeholders, and (6) refining the anti-counterfeiting strategy.

- *Withdrawing counterfeit articles from circulation.* Initiating product seizures is the first step to avert danger from the customer. For counterfeit goods that made it into the supply chain of a licit partner, brand owners can expect the responsible person (i.e. the shop manager or the logistics service provider) to be supportive with respect to both removing the illicit articles and providing information on how he or she obtained the products. However, if the holder of the illicit goods is consciously involved in the criminal activities, he is very likely do what is possible to deny access to the products. The brand owner may require search warrants and support from the police, and often the time that is necessary to initiate these steps is sufficient to remove the questionable articles. As a rule of thumb, seizures become more difficult the earlier they take place in the counterfeit actor's value chain. However, the closer they are to production, the more likely it is to cut off counterfeit supply.
- *Warning those who are affected by the imitations.* In case of imminent danger to the consumers, brand owners have to immediately inform the people and institutions concerned. Such a statement should describe how to recognize the counterfeits, what to do with them, what to do after the product has already

been consumed, whom to inform for example about the number of counterfeits confiscated, and what the owners of counterfeit goods can expect from the brand owners. The latter may consider compensating licit vendors who have counterfeit products in stock – they may otherwise be tempted to sell them to limit their personal disprofit.[26] The timeliness of information is critical, not only to prevent accidents but also to document that the company is properly dealing with the case. A generic "counterfeiting alert" should be prepared before an incident occurs to have it available "on call", alongside a press release, and, if necessary, a template for further information on corporate websites.

- *Finding/eradicating the source of the illicit products.* The occurrence of counterfeit goods may be seen as some sort of mishap, but it also offers the opportunity to prevent future incidents. Brand owners should approach it in a constructive way and collect as much information as possible to increase the chance of successful prosecution of offenders.

- *Seeking the prosecution of offenders.* Prosecution of offenders is the logical consequence of identifying illicit actors. It is, however, a difficult process. Especially in those countries that dominate counterfeit supply, the "big guys" behind counterfeit production are extremely difficult to catch. We nevertheless stress the importance of taking legal steps, even if they do not directly affect the illicit financiers. They may nevertheless lead to seizures of production machinery and unsettle middle men, which can dramatically increase the cost of counterfeiting.

- *Managing the relationship with the informant and other supportive stakeholders.* Most reactions are initiated after receiving hints from individuals such as customs officials or managers from retail stores who are not directly associated with the brand owner. Not following their advice or not providing any feedback is very likely perceived as a lack of interest on the part of the brand owner, and consequently can lead to an indifferent attitude of the informant when further counterfeit cases occur. This may significantly reduce the performance of future anti-counterfeiting efforts. As a consequence many successful companies follow every piece of evidence provided by important stakeholders to ensure their future support.

- *Refining the anti-counterfeiting strategy.* The last step emphasizes the importance of constantly refining the company's anti-counterfeiting efforts. Details from individual cases should be fed back to adjust the monitoring process and improve prevention techniques.

Reaction processes are, as are most complex business processes, highly dependent on the affected product, country, and the nature of the individual case. The

---

[26]However, reimbursement can also lead to a loss of motivation to carefully authenticate products when purchasing them or may even develop into a new business model of illicit actors who "sell" directly to the brand owner.

above-mentioned issues should nevertheless summarize the room for maneuver of affected brand owners.

5.3   Preventive measures

Strategies to fight counterfeit trade can be ascribed to one or more of the following categories: securing the company's supply chain, eliminating production of counterfeit products, hampering their distribution, discouraging or preventing users or consumers from purchasing faked goods, and limiting the damage that may result from illicit products. In order to achieve the corresponding goals, measures are required which are organizational, technological, legal, and communicative in nature (c.f. Figure 5.2). Companies usually choose a combination of these measures to tailor a mitigation approach which reflects their individual risk profile. The individual measures are described in greater detail in the remainder of the section.

Mitigation strategy	Organizational	Technological	Legal	Communicative
Securing the supply chain	++	++		
Eliminating illicit production		+	+	+[2]
Hampering illicit distribution		+	+	+[2]
Stopping deceptive consumption	+	++		+
Stopping non-deceptive consumption		+[1]	+	+

1) Only if security mechanisms can affect the functionality.
2) Only indirectly over legal measures.

Figure 5.2: Generic mitigation strategies to avert counterfeit trade

Organizational measures

We already highlighted the importance of organizational measures for monitoring and reaction processes. For prevention an adequate organizational design is of equal importance. It helps to secure the supply chain, hampers illicit distribution and deceptive consumption and limits loss or damage when illicit products eventually make it to the customer.

With respect to supply chain security, companies have to take care of two threat scenarios: first, non-original components that the company purchases itself (and that may end up in their products); and second, faked versions of the company's product which are traded as deceptive or non-deceptive counterfeits on the target market. For the first scenario a careful selection of component suppliers and sub-contractors is of particular importance. When purchasing goods companies should consider the risk imposed by completely counterfeit products as well the risk of substandard parts that have been built into otherwise genuine products. Especially in the latter case anti-counterfeiting technologies often fail to offer protection as they rarely can authenticate all individual parts or ingredients. Therefore manufac-turers have to rely on their suppliers – and should include counterfeit-related aspects in their auditing process. Anti-counterfeiting measures should also be reflected in purchasing strategies for B and C goods (for example power cords, batteries, etc.) since these categories are frequently targeted by illicit actors. Even if a supplier is well-known and trustworthy, its problem awareness has to be questioned.

To hamper the distribution of imitations, it is desirable to maintain tight control over the distribution channels. Several companies from within the luxury goods industry, for example, do not sell their products over the Internet, but make use of carefully selected networks of distributors in order to reduce the risk of faked goods being sold through trustworthy channels. Corresponding contracts allow for termination of the partnership if the distributor neglects its duty of care. Tightly controlled distribution networks greatly reduce the number of deceptive counter-feit cases in consumer markets.

Technological measures

Technological measures are an integral part of many anti-counterfeiting strategies. They serve as a means to authenticate genuine goods and thus help to distinguish them from counterfeits, or, for certain product categories, increase the production costs for illicit actors and confine the functionality of faked articles. Holograms, flip colors and micro printings are all prominent examples of established protec-tion mechanisms. However, copying these static features constitutes an ever-lower barrier for illicit actors, and many imitations today resemble their genuine coun-terparts so closely that their inspection becomes a time-consuming process. More secure features such as chemical and biological markers are often not suitable for large-scale testing – but in a market where an increasing number of counterfeit goods intermingle with mass-produced items, large samples or even complete checks are necessary. The latter factor also renders covert security features impractical, which, in this scenario, would require a large number of insiders to know about the characteristics of the hidden feature, which again would impose a security risk.

Another severe drawback of established anti-counterfeiting technologies is the limited ability or motivation of the user to check for the product's authenticity. Since security mechanisms are usually changed after being compromised, their users must be constantly kept up to date. Genuine products with different features often coexist until an old product line has been sold out, which complicates the checking process even more. As a consequence counterfeiters try to leverage the confusion among licit stakeholders (c.f. Section 8.2 for a thorough discussion of possible attack scenarios), which renders many technologies ineffective.

Emerging Radio Frequency Identification (RFID) technology constitutes a new approach to consolidate supply chain performance and security. Following the Food and Drug Administration's initiative to protect the U.S. drug supply, Pfizer, Purdue Pharma, GlaxoSmithKline, McKesson, CVS Pharmacy, and Johnson & Johnson among others have already piloted RFID-based track-and-trace solutions as a means to protect high-risk products such as Viagra or Oxycontin. Advantages over existing technologies are the possibility to automate large-scale tests, the ability to change underlying security protocols while maintaining the user interface, and the potential high level of security. High costs for the infrastructure, objections to data access and sharing, and privacy concerns among consumers are viewed alongside the technology's potential to avert counterfeiting, to achieve higher supply chain visibility, enhanced production, inventory and distribution control, and to implement efficient replenishment procedures. Chapter 8 covers RFID technology in detail.

Legal protection

Legal protection of intellectual property forms the basis of every anti-counterfeiting strategy, although legal measures are mostly regarded as a necessary, but by no means sufficient action. In any case a prompt and comprehensive registration of trademarks is a prerequisite to ban the production and distribution of counterfeit goods. Civil and criminal enforcement are potential legal measures to achieve injunctions against future infringements, compensation for damages, or statutory damages per type of product sold. However, counterfeit producers are mostly located in countries where intellectual property laws are difficult to enforce. In reality the right holders can only rarely eradicate the source of faked products. To improve the situation the Unites States and the European Union have recently launched wide-ranging collaborative initiatives to strengthen their industries against counterfeit trade. Public Private Partnerships and a stronger presence of coordinators in South East Asia and Eastern Europe aim to foster information exchange and encourage joint actions against the producers and distributors of fakes. Companies should take advantage of these programs. Customs, in particular, has proved to be a powerful ally if it is adequately supported by the right holder. Integrating customs in corporate anti-counterfeiting strategies requires formal applications for action in Europe or registration on the Principal Register of the Patent and Trademark Office in the U.S.

Communicating to the public

Companies which appear to be successful in protecting their brands and products regard public communications as the most promising individual anti-counterfeiting measure. In fact, communicating to the public is an effective way to influence both the supply- and demand-side of the counterfeit market. This can be done by increasing the awareness of the existence of counterfeit articles and by stressing the negative consequences of the consumption or usage of such goods. As a recent study revealed, a shift in the opinion of consumers can be most effectively achieved by

- stressing the uncertain quality of imitation products,
- emphasizing the personal risks for health and safety, outlining, for example, cases of dangerous chemical substances in clothes and children's toys, exploding batteries, unsafe car parts, etc.,
- outlining the embarrassment potential of non-deceptive fakes, and by
- highlighting the negative consequences for the job market and the national economy.[27]

Another objective of brand owners is to gain official support in the fight against counterfeit trade. They push their governments to put pressure on countries which are strong producers of fakes in order to force them to introduce and more strictly enforce intellectual property rights.

Info Box 5.2: Industry associations

There are many business communities that petition for greater commitments by national and international officials in the enforcement and protection of intellectual property rights. These interest groups also promote the information exchange between their member companies and aim at informing the public of the negative implications of counterfeit trade. Prominent associations include:

- The Business Action to Stop Counterfeiting and Piracy (BASCAP), www.iccwbo.org/bascap
- The Coalition for Intellectual Property Rights (CIRP), www.cipr.org
- The Comité Colbert, www.comitecolbert.com
- The European Communities Trademark Association (ECTA), www.ecta.org
- The European Brands Association (AIM), www.aim.be
- The International Anti-Counterfeiting Coalition (IACC), www.iacc.org
- The International Trademark Association's (INTA), www.inta.org

---

[27] Empirical evidence was found to support the first three statements in Europe, Japan, and North America, while the last statement (impact on job market and national economy) was only supported by a study conducted in the United States.

When communicating to the public, a campaign should not leave the impression that a given brand is especially prone to fakes and therefore more risky to use than the products of its competitors. Industry associations (see Info Box 5.2) can provide a suitable means of communication while keeping individual brands out of the discussion.

5.4 Organizational structure of anti-counterfeiting units

The organizational structure of a company's brand- and product-protection program has to reflect the different characteristics of counterfeit trade at the various stages of the value chain. Employees in the purchasing department, for example, may have (and may need) a different perspective on the problem than those in the legal, marketing or after-sales departments. Moreover, the way counterfeit goods are sold and the potential responses to such infringements vary from country to country and from product group to product group.

As is highlighted in the benchmarking study, a local presence is crucial for both monitoring and reaction measures. On-site employees are often the first to notice market breaks or to hear from dissatisfied customers. Moreover, good personal working relationships with external stakeholders are necessary for getting hints on counterfeit activities or extensive support from local enforcement agencies. Without good personal contacts, even licit vendors are less likely to inform brand owners of dubious offers they have received, and customs officers that have been turned down after reporting minor offences may be less eager to follow up on larger cases in the future.

Relying on a purely decentralized approach, however, is not always sufficient. Often local resources to deal with larger cases are not available, professional legal support is not at hand, and a consistent struggle against illicit actors may even come into conflict with other management objectives (local sales targets vs. time to invest in fighting counterfeits, local budget constraints vs. investments in the protection of the brand, etc.). Moreover, information on counterfeit occurrences should also support decision making on a company-wide basis and facilitate a transfer of knowledge among different business units.

Successful companies strike the balance between providing incentives for local action and cooperation on a company-wide basis. The organizational structure of their anti-counterfeiting programs mostly consists of a central staff unit that reports to the chief security officer or directly to the board (to highlight the engagement of top management), a task force with members from the different departments, and a network of local contacts throughout the firm. In Info Box 5.3 we outline a sample organizational chart of an anti-counterfeiting program.

Info Box 5.3: Organizational chart of an anti-counterfeiting task force

The organizational structure outlined below is an example of how large, multinational companies can set up their brand- and product-protection task force.

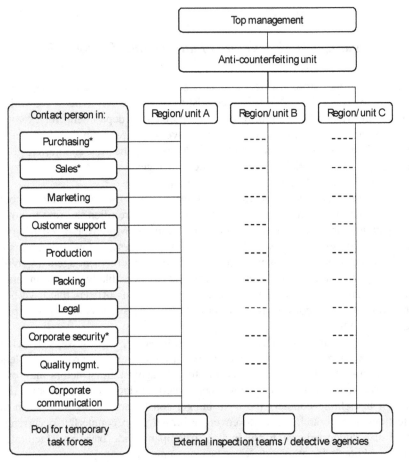

*) Suitable functions to head local monitoring activities

The central element is the anti-counterfeiting unit which defines the overall brand- and product-protection strategy, develops guidelines for monitoring activities and reaction measures, and decides upon the application of product-protection technologies. The anti-counterfeiting unit also provides legal support for local subsidiaries, develops training programs and awareness raising campaigns inside and outside the company, and collaborates with related international interest groups. The head of the anti-counterfeiting unit reports directly to the top management.

The top management's key responsibilities are budgeting and the evaluation of the program's success. Moreover, the senior executives should emphasize their commitment to the problem, e.g. by representing the company in an industry association that addresses intellectual property rights issues and by communicating the importance of anti-counterfeiting measures to other stakeholders.

A close cooperation with the different business units and local branches is facilitated by dedicated product and brand-protection mangers. They work together with contact persons within the departments, communicate the guidelines for monitoring and reporting, are active in regional industry associations, put together temporary task forces if larger counterfeit cases become known, and coordinate the work with external service providers that support product inspections, conduct test purchases, etc. Moreover, the product and brand-protection mangers support customs and involve the company's supply chain partners in the anti-counterfeiting measures.

Part C   Management Tools – Towards a Fact-based Managerial Approach

6 Determining the Market Share of Counterfeit Articles

Practitioners have to evaluate the effects of counterfeiting on individual brands or companies, conduct risk analyses, assess the performance of countermeasures or decide upon the allocations of funds to specific products or geographic markets. Estimates on the market share of illicit goods provide the basis for such considerations. However, obtaining quantitative data is difficult, and so is aggregating it into meaningful statistics. Existing market share estimates are hardly supported by reliable data, and most statistics appear to be biased towards unreasonably high figures. While many publications assume an overall market share of 7% of world merchandize trade, we found such estimates to reflect the situation only for a very small subset of brands and products, but by no means to be valid as an average share among merchandize trade. This discrepancy – as well as the need for brand-specific data, c.f. Info Box 6.1 – gave rise to the necessity of a transparent framework for estimating the counterfeit market share.

The remainder of this chapter is organized as follows. In the following section the background on counterfeit market share calculations will be provided by evaluating existing estimations. Thereafter a computational framework is introduced, and potential data sources are discussed. The framework is then exemplarily applied to the European (EU-25) and the U.S. American markets. Together with data on international merchandize trade, the findings make it possible to derive an upper boundary of the worldwide market share of counterfeit goods. The chapter concludes with a discussion on the validity of the results and the framework's applicability for individual brands and products.

Info Box 6.1: On the need for brand- and product-specific market share estimates

Even for the same brand, the market share of counterfeit goods can vary considerably among different product series. Illicit actors tend to focus on those goods which offer the highest price premium – in most cases the top of the line product – and quickly follow the brand owner when a more expensive series is launched. Low-cost versions and especially cheaper alternatives to well-known brands are less frequently targeted. Consequently, cross-industry market share estimates are at best a first indicator of the magnitude of the problem, but they should not make up for product-specific investigations. For the same reason, extrapolating data from market leaders in order to obtain the overall share of counterfeit goods often introduces a bias towards overly high estimates.

6.1   A critique of existing statistics

Estimates on the extent of counterfeit trade are frequently cited in the press (for example Business Week 2005), in industry white papers (for example ICC 2006), government reports (for example OECD 1998 and 2006), juristic recommendations (for example Leahy 2006), and academic publications (for example Gentry et al. 2006). As we found from a first review of existing literature, however, substantiated market share analyses are sparse. Furthermore, most studies appear to originate from industry associations which neither discuss the underlying data nor outline the applied methodology or computational rules.

CEBR. The study "Counting counterfeits: Defining a method to collect, analyze and compare data on counterfeiting and piracy in the Single Market," published by the European Commission and conducted by the Centre for Economics and Business Research Ltd (CEBR 2002), addresses this shortcoming. It aims to develop methods to support governments or private organizations in their effort to generate robust and comparable estimates of counterfeiting and piracy for various industries. The authors propose a so-called "methodology decision tree" to help users select an appropriate approach for data collection with respect to different products. Possible recommendations for data collection are to either use surveys of distributors/retailers, surveys of counterfeiters, mystery shopping and expert evidence, seizure data and detection rates, or consumer surveys. The study, however, misses guidelines on how to derive the estimates based on the input data.

Furthermore, the study discusses how much effort a company should expect to invest to collect the required data. However, the authors do not provide any justification for the recommended sample sizes. Their recommendation does not seem to reflect the difficulties associated with obtaining an unbiased sample, for example with respect to the selection of participants from different income groups and the identification of deceptive counterfeit consumers.

The model has its focus on consumption measures rather than on production estimates as consumption would be easier and less costly to estimate. However, our research found the opposite to be true. Like for estimates with respect to the illegal drug market, consumption data is largely missing and most difficult and costly to collect. As a recent UNODC report states, "the lack of this information has been one of the biggest constraints to market analysis on the demand-side and thus a main stumbling block to almost every attempt to gain greater insights into the market" (UNODC 2006: 126).

BSA/IDC. A second approach is provided by the Business Software Alliance (BSA) which – together with the International Data Group (IDC) – conducts worldwide estimates of the counterfeit software market on an annual basis. The underlying assumption of the study is that the quantity of pirated software equals total demand less total legitimate supply. Total demand is found using estimates on the average number of units of software installed on each PC and the number of systems in the market, total supply is derived based on software revenues and average software price. The calculation is performed for different categories, such as business and consumers applications, new and existing computers, and different software categories (BSA 2006).

The underlying assumption of the relationship between piracy, supply, and demand is highly questionable as it does not clearly separate piracy from other forms of substitution. Open source software, software transferred from old PCs etc., should be an integral part of the equation or at least be represented by a carefully derived corrective factor. The latter in fact is very difficult to estimate.

Input stems from consumer and industry surveys. It is not disclosed, making it difficult to reconstruct the analysis or to discuss the validity of data. The latter is especially important when the organizations providing the data or conducting the analysis potentially have an interest other than deriving correct estimates.

While the estimate on the share of counterfeit products ("35% of the software used is pirated") can neither be verified nor falsified given the nature of the study, the conclusions on the monetary volume (USD 35 billion) appear to be rather misleading. The BSA assumes that every pirated piece of software replaces one genuine product, which is unlikely especially in the emerging economies where free open source substitutes are available, and claims that decreasing piracy by 10% would add approximately USD 70 billion in tax revenues and an additional USD 400 billion of Gross Domestic Product (GDP) to local economies. It states furthermore the projected sector growth and the number of additional IT employees on a country level. The underlying assumptions are not made transparent, and neither are the data extrapolation methods nor the methodology to combine data from different regions explained; research findings for example on substitution effects and the impact of piracy on software diffusion are not taken into account (c.f. Givon et al. 1995).

IFPI. The International Federation of the Phonographic Industry (IFPI) aims to provide a more realistic picture of the counterfeit market (Jopling 2005 and IFPI 2006). It follows a mixed methodology approach based on consumer surveys, enforcement statistics (for example TAXUD 2005), test purchases, seizures of production facilities and the estimation of their production capacity. The share of counterfeit media is given in terms of the number of pirated articles (mostly CDs and DVDs) and the "physical piracy value". The latter, other than the BSA estimate, takes into account the discounted price for which counterfeit media are being sold. The study, however, does not outline the applied methodology in greater detail, relies on individual market estimates of their members in different countries, and does not state the underlying assumptions for compiling the overall estimates, which limits the study's transparency.

OECD/ICC. The frequently cited reports published by the OECD and the International Chamber of Commerce (ICC) are not based on independent calculations but combine individual studies from industry associations and lobbying groups (for example OECD 1998, ICC 1997 and 2006), and extrapolate the missing data. Estimates range from 5 to 7% (OECD 1998) to 7% (ICC 2006). As a recent report by the OECD (2006: 50) puts it, "the metrics underlying the ICC estimates are not clear. Some have interpreted the figure to mean that counterfeit products traded internationally account for 5 – 7% of total traded goods; others have indicated that the figure relates total counterfeit production (which would include production for domestic consumption as well as export) to world trade. Nor is it clear what types of IPR infringements are included in the estimate." In fact, the studies do not support a critical and unbiased assessment of the situation.

Evaluation and suggestions for improvement

We argue that the above-mentioned studies suffer from various methodological weaknesses that eventually lead to unreliable estimates of the extent of counterfeit trade. In detail, these shortcomings can be categorized as follows:

- *Unclear data sources.* Statistics rarely reveal the source of the underlying data, and are often based on reports published or collected by anti-counterfeiting associations with a strong interest in emphasizing the magnitude of the problem.
- *No precise definitions.* Frequently no stringent definition of counterfeit trade is applied. Authors often include other illicit activities such as factory overruns, parallel imports, smuggling or trade in stolen goods. Without a concise definition the degree of freedom for assembling market share estimates is large, which limits the explanatory power of such analyses.
- *Unrealistic assumptions on counterfeit prices.* Many calculations express the size of the counterfeit market in terms of monetary units without explaining how to translate the actual numbers of counterfeit products into their monetary equivalent. Using prices of genuine products may even lead to misleading statistics if not clearly stated.
- *Unclear relationship between revenue and earnings.* Some statistics mistake loss of revenue with loss of earnings, mostly without even being able to estimate the loss of revenue.
- *Unclear size of units.* Often the underlying data does not define the size of the counted units, i.e. does not explain whether one count is a single item or the unit of sales.
- *Incomplete sample and vague extrapolations.* Many statistics are based on a small and possibly biased sample. Often only the leading brand owners report seizures or counterfeiting-related data, which is then extrapolated to the overall market. This, however, does not take into account the concentration of counterfeit producers on the strongest brands.
- *No statement on error margins.* Given the potentially large errors of individual estimates, the specification of error margins is important for a meaningful

interpretation of the findings. However, most estimates do mention potential inaccuracies or even imply an unreasonably small margin.

- *Disclosed Methodologies.* Most authors do not describe the underlying methodology, making a validation or inter-study comparison difficult.

The conclusions for estimates of global counterfeit trade that can be drawn from our review are twofold. First, substantiated estimates should exhibit the following characteristics: they should (1) apply a concise definition of counterfeit trade, (2) use multiple, verifiable sources for the underlying data, including information on import and consumption, (3) allow for a comparison of the results over time, and (4) clearly outline the underlying methodologies and assumptions. Second, the use of multiple sources or techniques for obtaining the same information in various ways (triangulation) is essential as single sources often do not provide the required level of detail and may even be erroneous. Especially due to the clandestine nature of counterfeit trade, a triangulation approach helps to average out errors and to detect faults in the measurement and calculation process.

A framework to estimate the extent of counterfeit trade

The framework presented below aims to overcome the above-mentioned shortcomings. We follow an approach similar to the one used by the United Nations Office on Drugs and Crime (UNODC) for monitoring the illicit drug supply chain (Frate 2006 and UNODC 2006). The UNODC's frequently used and highly elaborated methodology is outlined in Info Box 6.2.

Info Box 6.2: Estimating the size of the illicit drug market

The UNODC model is based on the assumption that demand equals supply plus or minus changes in stock minus seizures and losses. It follows a mixed method approach to estimate each term of the equation. Demand is found as the product of the number of users and average quantities consumed per user, where the number of users is determined based on household and school surveys, multiplier methods (for example over surveys to determine the share of addicts who have been treated and the number of people in therapy), and capture-recapture methods (e.g. over the comparison of different registers such as treatment and arrest). Establishing and combining various methods helps to reduce the error margin. Supply of non-synthetic drugs, however, is easier to estimate than those of counterfeit products since satellite images to determine the sizes of plantations in combination with yield surveys provide rather precise data. A combination of supply and demand calculations refines the overall estimation. Not revealing the high level of uncertainty, the study clearly states the error margins of the result (USD 360 billion +178%; −76%) and thus allows for a careful interpretation of the findings.

Four principles guided the conception of the framework and the following calculation of our estimates. First, the methodology and the model were kept as simple as possible but reflect the basic characteristics of the counterfeit market. Second, the numbers of required assumptions is minimized and those which have to be made are made transparent. Third, the input data is to be available or can be generated with appropriate effort. Transparent assumptions and input data make it possible to calibrate the model when new information or findings become available. Error margins are specified. And fourth, the narrow definition of counterfeit goods is applied as developed in Section 1.3.

Our framework is based on a simple sink-source model, i.e. the primary assumption is that what is being produced, minus loss and seizures, is consumed. The starting point for the analysis is the primary sources of counterfeit products within a country or economic region, i.e. counterfeit imports as well as internal production or assembly (c.f. Figure 6.1). Sinks are exports, seizures, and consumption. Scrap and unsaleable articles are neglected. Buffering effects are assumed to be compensated over time, i.e. we assume that stocks which may build up or be depleted only distort the estimate in the short term. Under the assumption that the system is self-contained and buffering effects do not apply, the consumption of counterfeit goods within the market under study C equals the sum of imports I plus internal sources of counterfeit production P, i.e. production within the market under study, minus seizures S and exports E:

$$C = I + P - S - E \qquad (6.1)$$

In the following, we first provide a brief discussion of potential data sources for our model and continue with the details of calculating its individual components.

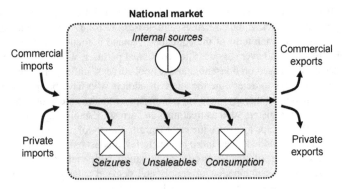

Figure 6.1: Structure of the sink-source model

Data sources

Valuable data sources for estimating counterfeit trade are enforcement statistics, consumer surveys, sampling (for example test purchases or inspections of warehouses), company data, indirect measures, and data on counterfeit production facilities, each with specific advantages and disadvantages which are briefly discussed below.

Enforcement statistics. Border inspections lead to approximately 70% of all seizures in North America and Europe, making customs one of the most powerful stakeholders with deep insights in the illicit market. With an inspection rate of 3 to 6%, the sample size is high, leading to relatively reliable estimates at least for frequently counterfeited brands and products. Customs statistics exhibit the required granularity as they are product-type- and brand-specific. Moreover, customs adheres to a narrow definition of counterfeit trade according to the TRIPS agreement, excluding activities such as parallel trade or trade in stolen goods. The information is available at no cost for the right holder, and, in a more aggregated form, also to third parties. Customs and enforcement statistics can be used to assess counterfeit occurrences in terms of number of articles, but do not always provide sales prices of imitation products and thus are alone not sufficient to estimate the monetary volume of counterfeit imports.

Among the downsides of enforcement statistics are a potential bias due to profiling schemes during inspection, the changing priority given to intellectual property rights infringements, and the influence of right holders on the probability of seizures, for example by filing applications for actions, providing information on security features of genuine products, or even initiating targeted exams (Maricich 2005). The quality of enforcement statistics varies from country to country, and there are also variances in the interception rates. However, high-quality data is available for the European Union, the United States, and most other states with strictly enforced intellectual property rights.

Consumer surveys. Consumer surveys constitute a powerful tool for assessing consumer attitude, awareness and behavior such as purchasing frequency, value, and volume with respect to non-deceptive counterfeit goods. They are important for understanding substitution effects and consequently for expressing the counterfeit market share in terms of monetary units. Their explanatory power is mostly limited to those goods where the user is aware of the illicit nature of the article. Since the purchase of counterfeits is often regarded as socially questionable behavior, surveys are especially sensitive to their design. Care has to be taken such that the survey results are representative of the entire population. Despite the critical design issues and their usually high costs, consumer surveys are an important complement to enforcement statistics which often do not capture private, small quantity imports for example by tourists. Demand-side surveys, however, are not sufficient alone as they do not necessarily capture deceptive counterfeit cases.

Company surveys. Company-internal data, for example on regional sales figures and warranty cases, can provide information on counterfeit occurrences in the market under study. Especially for goods with highly volatile demand or slowly changing counterfeit market share, sales figures and deviations with regard to demand forecasts are only of limited explanatory power. Warranty cases, however, can lead to reliable estimates for deceptive counterfeit goods. As counterfeits are expected to be often of low quality, they are more likely to be found among product returns than genuine articles. Drawbacks of this approach are the dependence on the quality of the fakes and thus the unknown probability of consumers turning in illicit goods.

In general industry surveys are not sufficient alone as they do not represent non-deceptive counterfeiting that has been established in completely parallel licit markets (such that companies are not even aware of their existence or have no access to them at all). For macroeconomic studies, other problems result from the difficulties to get responses which reflect the structure among the various industries; often, mostly companies with a strong perception of counterfeit issues participate in surveys, which eventually results in biased data. Moreover, care has to be taken as brand-protection managers may deny the existence of the problem or may regard surveys as a communication and lobbying tool and therefore tend to provide estimates which are on the high side.

Sampling. Sampling connotes the investigation of individual products within the supply chain or at the points of sale. The approach is potentially powerful but expensive; investigators often have to purchase sample products since a visual inspection of the packaging is rarely sufficient to spot faked goods. Tests are often destructive so investigated products cannot be fed back in the supply chain. In the case of a diverse retail structure and low counterfeit market penetration, the sample size has to be high in order to allow for a generalization of the findings. To avoid a bias of the sampling data, examination must not focus on suspicious vendors only. Besides the data to support statistics on the market share of counterfeit goods, sampling also helps to obtain a better understanding of the illicit retail structure, the quality and sales prices of counterfeit articles.

Indirect measures. Some goods allow for measuring by-products of counterfeit trade and for estimating the actual market share from these observations. Measures of side effects may subsume the monitoring of sales of complementary products (as done with computer hardware, operating systems and word processors) or the health implications resulting from the consumption or use of counterfeit products; both approaches are practical for only a small subset of products but may allow for obtaining in-depth market insights. As an example, counting discarded products or packaging constitutes a reliable approach which is suited for both deceptive and non-deceptive counterfeit cases. In order to reduce the bias, pack or product collections should be conducted at a number of carefully selected locations. The approach is expensive as it may require time-consuming investigations, and long lifetimes of individual products may further complicate the valuation. However, pack collections allow for calibrating other observation techniques. It may help to determine the success factor of customs inspections or to evaluate the validity of consumer surveys (c.f. Info Box 6.3).

Data on counterfeit production facilities. Information on production plants, seized production machinery or cash flows of counterfeit producers constitutes another potential source of data to derive the market share of counterfeit goods. However, since the number of producers and their characteristics are unknown, findings on the supply-side are difficult to generalize and thus often only of limited use.

Info Box 6.3: The TMA's approach to estimate the counterfeit market share

Some goods allow for measuring by-products of counterfeit trade and for deriving the actual market share from these observations. A prominent example is the analysis of discarded packaging materials. The Tobacco Manufacturers' Association (TMA), for example, estimates the prevalence of counterfeit cigarettes in the U.K. by collecting of discarded cigarette packs. Results produced by the TMA in June 2005 imply a total counterfeit penetration in the UK market of 2% to 3% (Bourn 2005). Such campaigns supply valuable information and help to calibrate other, less expensive approaches e.g. based on an analysis of customs statistics. However, they require the locations where the collections take place to be carefully chosen in order to avoid biased estimates. Moreover, conducting them is only feasible if the product under study is consumed in large quantities.

Detailed framework

In the following we present the structure of our framework proposal and describe its input parameters. Furthermore, we discuss options for quantifying each parameter based on the above-mentioned data sources.

Counterfeit imports (I) can be broken down into commercial and private imports, with commercial imports ($I_{Commercial}$) connoting the flow of goods organized by illicit actors which intend to sell the goods in the market of destination, whereas private imports ($I_{Private}$) subsume imports in small quantities, mainly by tourists for captive use or dissemination to friends.

In most countries with strictly enforced intellectual property rights, customs statistics constitute a reliable estimator for commercial imports. Based on the information on the quantity of seized articles ($S_{Customs}$), one can estimate the volume of imported goods with further knowledge on the share of inspected articles ($s_{Inspected}$), and knowledge on the selection bias and on the ability to identify counterfeits if they have been selected for an inspection ($s_{Success}$):

$$I_{Commercial,Articles} = S_{Customs,Articles} / (S_{Inspected} * s_{Success}) \qquad (6.2)$$

The volume of seized articles may be available in terms of the number of items, their estimated street price, or the corresponding value of genuine goods. While each unit of measurement has its advantages – unit counts for individual brands or products are not biased by additional considerations on their sales price, the street price aims to reflect the actual market price but allows for a high degree of freedom on behalf of the enforcement agency, and the corresponding value of genuine goods can serve to estimate an upper limit – it has to be stated which measure applies. The share of inspected goods is provided upon request by most customs organizations. Though the success rate ($s_{Success}$) is difficult to estimate, it constitutes a key performance measure of customs and thus an important element of their risk management assessments. It may nevertheless be a weak point of the overall estimation as the underlying data for deriving $s_{Success}$ is not necessarily disclosed.

Private counterfeit imports may constitute another noteworthy source of counterfeit articles. The phenomenon is sometimes referred to as ant-traffic and appears to exist particularly for non-deceptive counterfeit cases in countries where purchasing illicit products is difficult and where consumers do not fear prosecution when importing small quantities of fakes. The flow of goods may be best estimated based on an anonymous, paper-based consumer survey asking respondents for purchases of unreasonably cheap branded goods abroad during a defined period of time. For countries where counterfeit articles are not confiscated by customs if imported in small quantities, the overall volume of counterfeit imports is obtained by a summation of both estimates.

Internal production (P) describes the manufacturing of counterfeit goods within the country under study. Estimating production volumes directly is extremely difficult since illicit actors aim to carefully disguise their activities. However, if a brand owner succeeds in identifying a counterfeit production site and furthermore gets access to the machinery, the company can draw several important conclusions. In markets with strong illicit producers, capture-recapture models constitute a starting point for an analysis. When a production facility is being closed down, it is possible to estimate its output. Often it is also feasible to attribute counterfeit products which have been seized independently from the close-down to the given production facility. Knowing the share of imitation products which can be attributed to this facility, it is possible to draw conclusions on the overall production volume (c.f. Info Box 6.4). Similar approaches are used to estimate the number of drug addicts by comparing arrest registers and treatment registers.

Info Box 6.4: Capture-recapture methods

Capture-recapture models base on simple probability considerations which can be undertaken if a number of entries appear in two independent registers. Such methods are often used to estimate the size of animal populations, where researches "capture" and mark a number of animals, release them, and determine the share of marked creatures that are "recaptured" in the next season. The approach is also applicable to counterfeit market share estimates. If, for example, an illicit manufacturing site is closed down and the brand owner has the opportunity to investigate the production facilities, it is often possible to determine the number of counterfeit articles that have been manufactured within a defined period of time. It may also be possible to figure out how many of these articles have been confiscated by customs. Let us assume that 50.000 items have been produced within six months. Let us further assume that among the 100.000 products which have been confiscated 2.500 articles (i.e. 2.500/100.000 = 2,5%) came from this production site. Thus, it can be concluded that total counterfeit production is around 50.000/2,5% = 2.000.000 articles per half year. The method helps to establish additional estimates without extensive field research. The results, however, are biased if the seizures rates for the articles from the closed production site are higher or lower than the average seizure rates. Establishing various estimates (e.g. based on other closed down fabs, warranty claims, etc.) nevertheless helps to average out errors and thus to obtain better market data.

Due to the large number of different counterfeit products and affected industries, estimates on counterfeit production rely on company input. The overall estimates should be derived based on industry-specific data which is to be weighted by the corresponding revenue of each group, and industry-specific estimates also have to reflect the distribution of company sizes (c.f. Equation 6.3 and 6.4). As major brands are frequently affected disproportionately and often overrepresented in surveys, a simple extrapolation is not suitable.

$$estimate\_overall = \frac{\sum_i estimate\_industry\_i * revenue\_industry\_i}{\sum_i revenue\_industry\_i} \quad (6.3)$$

$$estimate\_industry\_i = \frac{\sum_{j(i)} estimate\_company\_i,j * revenue\_cluster\_i,j}{\sum_{j(i)} revenue\_cluster\_i,j} \quad (6.4)$$

Seizures (S) reduce the number of counterfeit articles which are available for consumption and thus constitute a sink for illicit imitation products. Most seizures are captured by enforcement statistics. The information can be complemented by company surveys which are almost always involved or at least informed when articles of their brand are confiscated.

Counterfeit exports (E) constitute an important term for those countries which are strong counterfeit producers. The reasoning for its calculation resembles the approach for estimating counterfeit imports.

Counterfeit consumption (C) can be expressed by solving Equation 6.1 for *C*. However, in order to increase the precision of the estimate, it may be desirable to also calculate *C* separately. Several approaches can be applied, for example consumer surveys or pack collections as outlined above.

6.2   Macroeconomic calculations

In the following we provide two examples for estimating the volume of counterfeit trade. For this purpose our computational framework is applied to the European Common Market (i.e. EU-25) and the U.S. American market. For both regions comprehensive enforcement statistics are available; the data is complemented by interviews with brand-protection experts from the luxury, fast-moving consumer goods, tobacco, and engineering industries as well as by a survey among German-speaking consumers. The analyses provide valuable insights which substantiate the criticism on the established estimates and highlight some noteworthy properties of such calculations (c.f. Info Box 6.5). They moreover provide guidance for companies which want to conduct similar analyses for individual brands and product groups.

Info Box 6.5: Noteworthy properties of the calculation

We will encounter several noteworthy issues during the calculation. First, we will see much higher margins of error than we are used to in other financial analyses. They stem from associated uncertainties of many underlying approximations. Second, we will discover that it is essential to make these assumptions as transparent as possible to evaluate the validity of the results. A reader may not agree with one number or another – so he or she should be able to adjust it, for example when future statistics or surveys provide additional input. And third, we will stress that error margins are not an imperfection of a calculation, but rather should be part of any substantiated analysis. Without specifying the accuracy of an estimate, its interpretation is often misleading.

Counterfeit imports into the EU-25

Following the computational framework, the sources and sinks of counterfeit articles are evaluated and combined according to Equation 6.1. We derive the estimate for counterfeit consumption within the EU-25 for the reference year 2005.

Counterfeit imports (I) are divided into commercial and private imports. Commercial imports are estimated according to Equation 6.5, with the parameters taken from enforcement statistics of European customs and interviews with senior customs officials. At the borders of the EU-25, 75.7 million counterfeit articles have been seized ($S_{Articles, Customs}$) (TAXUD 2006). According to talks with custom officials, the share of inspected goods lies between 3.5% and 4.5% ($s_{Inspected}$). The success rate ($s_{Success}$), i.e. the ability to select suspicious consignments, is assumed to equal 1.5 with an error margin of 50%.[28] This leads to interception rates for counterfeit goods between 2.63% to 10.25%. Equation 6.5 denotes the number of commercial counterfeit imports brought into the EU-25.

$$I_{Commercial, Articles} = \frac{75.70 \text{ m}}{0.04 \ [\pm \ 0.005] * 1.5 \ [\pm \ 0.75]} = 1.26 \text{ bn } [+129\%; -41\%] \quad (6.5)$$

For the subset of 15.5 million articles which have been confiscated by German customs, the equivalent value of the corresponding genuine items has been determined by the German Federal Customs Administration. Under the assumption that counterfeit articles yield the same sales prices as their genuine counterparts, their overall value was found to be EUR 213 million, with an average value per article of EUR 13.80 (TAXUD 2005 and Bundesministerium der Finanzen 2006). Since the underlying evaluation scheme does not account for discounted sales prices of illicit goods (imitation products are often sold as non-deceptive counterfeits where consumers expect considerable price deductions), we treat the value as an upper boundary for the further calculations. The estimated average street price of counterfeit articles of EUR 4.20, which was provided by German customs in 1999, serves as the lower boundary (Bundesministerium der Finanzen 2002 and TAXUD 2001). The volume of commercial imports is given in Equation 6.6.

$$I_{Commercial, EUR} = I_{Commercial, Articles} * \text{EUR } 8.54 \ [\pm 62\%] = \text{EUR } 10.78 \text{ bn } [+269\%; -77\%]$$
$$(6.6)$$

The volume of private imports has been investigated based on a survey among 203 German-speaking consumers above the age of 13. 26% of the respondents indicated that they had purchased "counterfeit or unreasonably cheap branded articles"

---

[28] A success rate greater than 1.0 represents the experience of customs officers and the benefits of heuristics that help to select high-risk products, for example based on the country of origin.

outside the EU within the last three years; among those, the average number of imported counterfeits was found to be 2.6 articles. For the German market with a population of 71.5 million citizens above the age of 13 (Statistisches Bundesamt Deutschland 2006), this leads to an annual private import of 16.1 million counterfeit articles; with an average price per privately imported counterfeit article of EUR 19.20,[29] this leads to a market volume of EUR 309 million.

An extrapolation to describe the flow of counterfeit imports into the entire EU-25 leads to a volume of private imports of EUR 1.74 billion. Given the diversity of buying power and travel activities within the European common market, the validity of this step is questionable. We therefore concede a wide error margin of 50%.

The overall influx of counterfeit imports is the sum of commercial and private imports (c.f. Equation 6.7 and Equation 6.8).

$$I_{Articles} = 1.35 \text{ bn } [+123\%; -41\%] \tag{6.7}$$

$$I_{EUR} = \text{EUR } 12.52 \text{ bn } [+239\%; -73\%] \tag{6.8}$$

Internal production (P) The European Union has strong protection mechanisms for intellectual property and efficient means to combat infringements. In fact, internal production appears to be only a minor source of counterfeit products. According to interviews with practitioners from the luxury goods, fast-moving consumer goods, tobacco and clothing industries, between 1% and 10% of counterfeit products (i.e. commercial imports) which have been traced back to the production facility come from one of the member states of the EU-25:

$$P_{Articles} = 69.40 \text{ m } [+316\%; -89\%] \tag{6.9}$$

$$P_{EUR} = \text{EUR } 593.00 \text{ m } [+571\%; -96\%] \tag{6.10}$$

Seizures (S). European customs claims to be responsible for between 70 and 90% of all seizures conducted within its territory.[30] With 75.7 million confiscated goods and average prices as used in Equation 6.6, this leads to the following volume of confiscated goods:

---

[29] Please note that the average price per private counterfeit import is based only on a convenience sample of 62 people. A more thorough investigation is required to achieve smaller error margins.

[30] Source: Interview with C. Zimmermann, Head of Sector, Counterfeiting & Piracy, DG Taxuds, European Commission, November 2006, Brussels

$$S_{Articles} = 94.60 \text{ m } [+14\%; -11\%] \qquad (6.11)$$

$$S_{EUR} = \text{EUR } 808.00 \text{ m } [+85\%; -66\%] \qquad (6.12)$$

Counterfeit exports (E) only play a minor role given the small production capacities of illicit manufacturers in the market under study. This is in line with enforcement statistics of potential target markets, which do not show significant counterfeit imports coming from the EU-25 (c.f. U.S. Customs and Border Protection 2005). Counterfeit exports are therefore neglected in this estimation.

Counterfeit consumption (C) is derived by solving Equation 6.1 for *C*. The wide error margin mainly results from unclear prices of counterfeit goods; the span includes both extremes as it captures low street prices of mostly non-deceptive counterfeit goods up to the assumption that the imitation products yield prices which equal those of their genuine counterparts:

$$C_{Articles} = 1.33 \text{ bn } [+143\%; -48\%] \qquad (6.13)$$

$$C_{EUR} = \text{EUR } 12.31 \text{ bn } [+275\%; -85\%] \qquad (6.14)$$

Counterfeit consumption amounts to 0.36% (0.05% to 1.33%) of merchandize imports of the market under study; on average, a European consumer knowingly or unknowingly spends EUR 26.60 (EUR 4.00 to 99.20) on counterfeit goods per year. Please note that we did not calculate point estimates but an interval of possible values for the counterfeit market volume.

Counterfeit imports into the U.S. American market

As for the European market, comprehensive enforcement statistics are available for the United States. However, due to a lack of additional data, we base our estimation solely on the seizure statistics provided by U.S. customs to derive the volume of commercial counterfeit imports.

Counterfeit imports (I). U.S. Border Control seized counterfeit goods with a domestic value of USD 93.23 million in 2005 (U.S. Customs and Border Protection 2005). Conceding an error margin of 50% for the customs estimate on the monetary value of the confiscated goods, and using values for $s_{Inspected}$ and $s_{Success}$ of 4.0% (± 0.5 percent points) and 1.25 (± 0.75), i.e. interception rates for counterfeit goods between 1.75 to 9.0%, results in the following estimated import volume:

$$I_{Commercial,USD} = \frac{\text{USD } 93.23 \text{ m } [\pm 50\%]}{0.04 \, [\pm 0.005] \times 1.25 \, [\pm 0.75]} = \text{USD } 1.87 \text{ bn } [+329\%; -72\%]$$

$$(6.15)$$

During this study no thorough consumer survey has been conducted to explicitly investigate private counterfeit imports of U.S. citizens. Given the similar characteristics of the intellectual property landscape, we therefore use the data from the German market survey, i.e. we assume that those who import imitation products on average purchase 2.6 articles for an average sales price of EUR 19.20 (USD 22.97).[31] With 16.41 million travelers visiting non-US and non-EU destinations (U.S. Dept. of Commerce 2005), goods worth USD 489.96 million (±50%) are imported privately. This results in an overall counterfeit import of

$$I_{USD} = \text{USD } 2.34 \text{ bn } [+271\%; -68\%] \qquad (6.16)$$

Internal production (P). As for the European Union, internal production is only a minor source of illicit imitation products. Only between 1% and 10% of the seized articles stem from production facility from within the U.S.:

$$P_{USD} = \text{USD } 103.00 \text{ m } [+679\%; -95\%] \qquad (6.17)$$

Seizures (S). The volume of counterfeit articles which are being confiscated and thus not entering consumption is found based on the same assumption which has served to calculate the seizures for the European market:

$$S_{USD} = \text{USD } 117.00 \text{ m } [+71\%; -56\%] \qquad (6.18)$$

Counterfeit exports (E). Enforcement statistics of potential target markets do not show a significant volume of counterfeit imports coming from the U.S. market (c.f. TAXUD 2005). Counterfeit exports are therefore neglected in this estimation.

Counterfeit consumption (C). The overall consumption is found to equal a share of 0.14% (0.03% to 0.55%) of the merchandize imported into this region.

$$C_{USD} = \text{USD } 2.34 \text{ bn } [+305\%; -76\%] \qquad (6.19)$$

Again, please note that we did not calculate the point value of the counterfeit market share but an interval of possible values. For the European market we found that between 0.05% and 1.33% of all goods consumed (in terms of value), and for the U.S. American market between 0.03% and 0.55% of all goods consumed are

---

[31] The assumptions, however, have been briefly validated in a non-representative paper-based survey in Boston, MA, in January 2007.

of counterfeit origin. We know that the error margin is relatively wide. However, the estimate is good enough to refute many disproportionate estimates, as we will see below.

Estimating the share of counterfeit goods among world trade

The EU-25, the United States and Japan[32] together generate more than 50% of merchandize exports and more than 60% of imports (WTO 2006). Given their pro-minent role, knowledge on the extent of imitation products in these markets allows meaningful conclusions on the share among world merchandize trade to be drawn. To facilitate the overall estimation, the world market is divided into four regions (c.f. Figure 6.2). Import volumes and the respective share of counterfeits are known for the EU-25, the United States and Japan, where the upper limits of the estimates are used in order to derive an upper boundary of the overall market share. Imports into the countries subsumed under "Rest of World" which originate from the EU-25, the United States and Japan are assumed to contain an unrea-sonably high share of counterfeit articles (i.e. 0.5% of their value) in order to ensure that the result constitutes an upper limit. For the remaining imports into the "Rest of the World" (including imports originating from other countries within this group) no reliable estimates are at hand. Therefore the share is treated as a parameter with values between 5% and 30%. The results are summarized in Figure 6.3. Again, we calculate the upper limit, not the most likely market share.

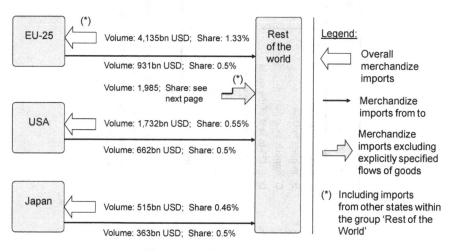

Figure 6.2: Schematic market model

[32] Population above 15 years: 110,193,000 (Japanese Ministry of Internal Affairs and Communi-cation 2007); World Merchandize Imports: USD 515 billion (WTO 2006); counterfeit goods seized: 1,040,000 articles in 2004, (±50%) (Dubois 2006); inspection rates: 0.04 (±0.005); suc-cess rate: 1.25 (±0.75); for the remaining parameters see EU-25.

Flow of goods	World merchandize imports in billion USD	Value of counterfeit goods in billion USD	Share of counterfeit goods among Imports (a)				Remarks
A Imports to EU25	4.135,0	55,0	1,33%				(b)
B Imports to USA	1.732,0	9,5	0,55%				(b)
C Imports to Japan	515,0	2,4	0,46%				(b)
D EU25 to rest of the world	931,0	4,7	0,50%				(c)
E USA to rest of the world	662,2	3,3	0,50%				(c)
F Japan to rest of the world	363,2	1,8	0,50%				(c)
G Other to / within r.o.w.	1.985,4	198,5 (d)	5%	10%	20%	30%	(c)
Total	10.323,8	275,2 (d)	1,7%	2,7%	4,6%	6,5%	

Legend:
a) Upper limits rather than most likely estimates
b) World merchandize trade figures from (WTO 2006: Table III.2)
c) World merchandize trade figures from (WTO 2006: Table A.2)
d) For a share of counterfeit goods of 10%

Figure 6.3: Global extent of counterfeit trade - Upper boundaries[33]

---

[33] Again, please bear in mind that the table gives upper boundaries. The best guess of the overall counterfeit market share is most likely closer to 1% to 2%.

Even for an average counterfeit market share of 30% within the group "Rest of the World", the worldwide share is still below 7%. Taking into account that merchandize trade, especially in the countries under consideration, also includes a con-siderable share of raw materials, fuel, staple food and other goods which are only rarely protected by trademarks or designs, it is unlikely to assume that 30% of all traded goods could be counterfeited. Moreover, the group "Rest of the World" also includes countries with strictly enforced intellectual property rights such as Canada, Switzerland, Australia, and Norway, which together make up more than one third of the trade volume in this group, and which have only been allocated to this group since no data on these countries has been at hand during this study. Following this argumentation, even a counterfeit share of 10% among this group appears to be a generous estimate, which still leads to an overall market share of below 3%.

In any case the frequently cited estimate of the OECD of 5% to 7% appears to be much too high. A share between 1% and 2% is a more realistic upper limit based on the previous considerations. Stating that, we certainly do not want to define down the problem – even a share of only 1% of mostly sub-standard imitation products can have severe implications for quality management and consequently for the relationship to the customer or even consumer health. Nevertheless, we believe that a more critical assessment of the counterfeit market share is important for a better understanding of the problem.

6.3 Microeconomic calculations

In the remainder of this section the computational framework is applied to two specific brands from the luxury and fast-moving consumer goods sectors. The analysis is based on publicly available information only; no internal-company data was taken into account.

Example A: Counterfeit trade in the luxury consumer goods industry

Brand owner and manufacturer A produces well-known exclusive, high-quality writing instruments. In the past few years the product range has been expanded to a wider range of luxury goods, including writing accessories, luxury leather goods such as handbags and belts, jewelry and watches. The principle design of many product lines has remained almost unchanged over the last few decades, and the company emphasizes tradition and craftsmanship over fashion trends and pop-culture. Characteristic logos are placed on every product, but do not dominate the design of individual articles. The brand is positioned globally and the company maintains operations in more than 70 countries. Assembly of the product line under study takes place solely in Europe and the retail structure is organized over company-owned boutiques and licensed partners. Neither the company nor its sales partners distribute products over the Internet as this is considered to be a sales channel that is difficult to protect. The following analysis relates to counterfeit cases of goods sold under the brand of manufacturer A within the EU-25 member states in the year 2004. It is based on publicly available information and on a consumer survey.

Commercial imports (I_{Com}). In 2004 European customs seized 203,000 counterfeit articles of the product category and brand under study. Seizures took place at the borders of or within the EU-25 member states. On average, 4.0% (±0.5 percent points) of all imported consignments were inspected. According to interviews with customs officials, inspection rationales with respect to the frequently targeted product usually yield high success rates of customs in identifying counterfeit goods during the inspection process, with values of $s_{Success}$ as high as 2.5 (±0.75). Using Equation 6.2, the number of commercially imported counterfeit articles is found to be 2.03 million (+63%;–32%).

Private imports (I_{Priv}). Data on private imports is available for the German market only. The consumer survey and its extrapolation to the European market indicate that a total of approximately 190,000 (+125%;–67%) counterfeit articles of the product under study are brought to the EU-25 per year.

Internal production (P). According to product and brand-protection experts from manufacturer A, production of counterfeit articles within the European Union only constitutes a minor source for the product and can be disregarded in this calculation.

Consumption (C). The analysis of counterfeit consumption can be conducted based on information on warranty claims and on data from extended, product-specific consumer surveys. Company-internal information, however, is not disclosed for publication, and consumer surveys require a larger sample size to allow meaningful conclusions to be drawn. The latter is especially true due to the small probability of a consumer actually having purchased a counterfeit article of the specific brand under study. For this reason consumer surveys to validate the findings were not included in this calculation.

Commercial and private exports (E_{Com} and E_{Priv}). Domestic production does not constitute a significant source of counterfeit goods within the markets under study. Therefore commercial and private exports can be disregarded.

Seizures (S). On average, customs is responsible for 70% to 90% of all seizures of counterfeit articles. Applying this estimation to the seizure statistics implies that 254,000 (+11%;–14%) articles were confiscated.

Summary of the findings. The overall number of counterfeit articles imported into the member states of the EU-25 is estimated at 1.97 million (+79%;–41%) items. The quantity is comparable to the estimated annual production of genuine articles. Further investigation revealed that approximately 90% of these fakes are poorly manufactured low-cost articles which are unlikely to pass as original products, and only about 10% could potentially be sold as genuine products. These findings accentuate the need for a collection of other counterfeit-related characteristics such as quality measures and sales prices alongside the data on seizure quantities. However, even without such additional data, the results provide a good baseline to evaluate the magnitude of the problem.

Example B: Counterfeit trade in the fast-moving consumer goods industry

Brand owner and manufacturer B is a major player in the primary alkaline battery market with a market share of above 40%. The brand stands for high quality, and the products are higher priced than most of the competitors. Within the product category under study, B sells approximately 750 million articles per year to the European market (EU-25) alone. The counterfeit market share of the product under study is estimated as follows.

Commercial imports (I_{Com}). In 2004 European customs seized 1.69 million counterfeit articles of the product under study at the borders of or within the EU-25 member states. On average, 4.0% (±0.5 percent points) of all consignments were inspected. According to interviews with customs officials, inspection rationales with respect to the frequently counterfeit products lead to high success rates for identifying counterfeit goods during the inspection process, with values of $s_{Success}$ as high as 2.5 (±0.75). According to Equation 6.2, the number of commercially imported counterfeit articles was found to be 16.93 million (+27%;–12%).

Private imports (I_{Priv}). The brand name under study is seen as a sign of quality and a means to reduce search costs. It does not communicate interpersonal values. No significant share of consumers would intentionally purchase counterfeit versions but choose cheaper genuine products instead. We can therefore neglect private imports.

Other sources. Again, the production of counterfeit articles within the European Union only constitutes a minor source for the product under study. Therefore it can be left out of this calculation.

Seizures (S). Customs is assumed to be responsible for 70% (±10 percent points) of all seizures of counterfeits of this brand, implying that 2.42 million (+17%;–13%) articles were confiscated.

Summary of the findings. Based on the preceding considerations, the overall number of counterfeit articles imported into the member states of the EU-25 is estimated at 14.51 million (+29%;–12%) pieces, with an approximate market share of 1.9% (+0.6;–0.2 percent points). Even when solely considering easily accessible, publicly available information, the analysis provides a reasonable estimate of the size of the illicit market.

7   Implications for Affected Enterprises

The far-reaching implications of counterfeit trade are widely recognized, and most brand- and product-protection experts are aware of the major risks for their business. However, detailed knowledge on the numerous long and short-term effects appears to be sparse, and many related questions have not been thoroughly addressed yet: Counterfeit trade can reduce revenue – but to what extent? Illicit imitation products can damage a brand name – but how do they interfere with individual performance measures of trademarks, such as name recognition, perceived exclusiveness, and quality associations, and what is the effect on brand value? Counterfeiters may benefit from learning effects during production and may turn into licit competitors in the future – but what about the resulting barriers of entry for contemporary licit competitors in emerging markets?

In the following chapter we will provide answers to these questions. We will investigate the potential loss of revenue due to short-term substitution effects, thoroughly analyze the impact on brand value, and introduce a set of tools to quantify financial losses that stem from these effects. Moreover, we will discuss the impact of imitation products on total cost of quality, liability claims and future competition. We will furthermore outline the potentially positive implication for example on network effects, access to a future user base, and product launches. Addressing these issues not only provides a substantiated basis to vindicate investments in anti-counterfeiting measures, it also helps a great deal to develop targeted protection strategies.

7.1   *Quantifying the loss of revenue*

Consumers may accidentally purchase counterfeit articles or knowingly buy cheaper imitations as alternatives to genuine products. The direct loss of revenue that licit manufacturers face when counterfeit articles compete with their original goods is probably the most obvious implication of counterfeit trade. However, not every faked article sold ultimately constitutes a lost sale among licit products, although many impact analyses are based on this assumption.[34] In fact, the extent to which a substitution takes place depends on numerous factors. These include the characteristics of the imitation product (product category, price, visual quality, etc.), the properties of the genuine good (price, function of the brand, expected level of quality and service, warranty, etc.), the seller (seriousness, appearance, place of transaction, etc.), and the consumer (intention, awareness, risk-taking, income, brand perception, etc.).

---

[34] See, for example, Bundesministerium der Finanzen (2006).

A good understanding of the substitution effects is crucial to conceptualize the concurrence of licit and illicit markets, as well as to estimate the loss of revenue among brand owners.

During our numerous industry projects, hardly any brand-protection experts were able to provide even a rough estimate of the financial implications for their business (a discussion on the need to quantify the financial impact is provided in Info Box 7.1). In fact, investigations of consumer choice regarding genuine and counterfeit products constitute a very young field of research. Until now no methodology has been published to analyze related substitution effects, while the corresponding assumptions among practitioners appear to be rather crude. Here, two contrary opinions seem to be dominant. The first group argues that counterfeit goods are sold into completely different market segments and thus do not interfere with trade in genuine branded goods, whereas the second group claims that every faked article leads to a corresponding loss of sales among genuine products. Both assumptions may be valid under very specific circumstances, for example for extremely low-cost knockoffs (zero substitution) or for counterfeit medicine in wealthy countries (complete substitution). However, neither consideration constitutes a realistic assumption for a wider range of goods as imitation products are partly sold as deceptive *and* non-deceptive counterfeits, exhibit a wide variety of quality levels, and often not only compete with well-known branded products, but also with generic merchandize.

We aim below to improve the understanding of the coexistence of markets with genuine and counterfeit goods. After providing the general background on consumer buying behavior and substitution effects, an empirical analysis of consumer

---

Info Box 7.1: **Different perspectives on the need to determine the financial impact of counterfeit trade**

While monetary valuations may be of limited interest in industries where counterfeits endanger the health and safety of consumers, such estimates can substantially support decision making on countermeasures when large numbers of imitations lead to a continuous loss of profit. Products such as fashion clothing and accessories, many fast-moving consumer goods, commercially counterfeited digital media, tobacco, etc., all fall under the latter category. Here, expenses of brand- and product-protection measures should be justified by return on investment considerations. While marketing experts and senior management were in most cases highly interested in related quantitative data, our effort to develop the corresponding calculation tool found only limited support among brand-protection experts. Among them the prevailing opinion could be summarized as "we should not waste time on such calculations but use it to go after illicit actors". However, we think that a quantitative approach will benefit companies as it will support investment decisions.

choice with respect to counterfeit articles is introduced. The findings allow a substitution factor to be estimated, based upon which individual companies can derive – or at least estimate with higher precision – the direct loss of revenue resulting from counterfeit trade.[35]

Background on consumer purchasing behavior and substitution effects

Knowledge of consumer choices regarding the selection and consumption of goods is of great importance for marketers and policy makers. Marketing researchers have been very actively engaged in studying consumer purchasing decision making for over 40 years. Various approaches have been published to describe, explain, and predict buying behavior, for example arguing for bounded rationality in the related decision-making processes (Conlisk 1996) and constructivist preference formation (Gregory et al. 1993), or proposing and discussing empirical models of consumer choice (Morrison and Schmittlein 1988 and Uncles et al. 2005). The field of research is likely to remain one of great interest given the technological development of products, new means for accessing information (for example Internet access via mobile devices or at the point of sale), the changing marketplace (for example online shops, online auctions), and other general trends in society (growing environmental concerns, and changing attitudes towards brands, etc.), which are all likely to alter buying behavior (Bettman et al. 1998). Questions of interest are, among many others, why consumers make the purchases that they make, what factors influence them, and how individual factors can contribute to a change.

The inclusion of counterfeit trade in these considerations constitutes an interesting and important field of research. We already outlined the fundamentals of consumer choice in such markets in Chapter 3. Here the focus is on the quantitative assessment of substitution effects among genuine and counterfeit goods. In other words we will try to determine the probability that a consumer would have bought a genuine product instead of the counterfeit for the following two cases. Firstly for non-deceptive counterfeiting, if the illicit good had not been available, and secondly for deceptive counterfeiting, if the illicit nature of the product had been known to the consumer prior to the purchase.

In this context the term *substitution* refers to one article displacing another, rather than to the price effects on quantity demand in macroeconomics (c.f. Hicks 1970). The substitution factor S is defined as the probability of one counterfeit article displacing a genuine item of a certain type and brand. If $S = 1$, for example, a consumer who purchased one counterfeit product would have bought one original product instead if he or she had known that the article was a counterfeit or if no imitation had been available. $S = 0$ represents the assumption that counterfeit

---

[35] The indirect, or long-term, effect on revenue that may result from changes in the perceived exclusiveness or expected level of quality is captured in the following section on the impact on brand value.

trade does not interfere with licit trade and thus does not lead to a loss of sale for the brand owner.

Empirical analysis of consumer choice with respect to counterfeit articles

For the investigation of consumer choice in markets with counterfeit goods we simulated a purchasing environment with and without imitation products.[36] Within this study 203 consumers were asked to assume they were planning to spend a given amount of money on a product of a defined category. In the first round they were given three options, namely to either purchase a counterfeit article, a product from a lesser known brand, or the original equivalent of the imitation product. Each article was characterized according to the associations which had been mentioned most often in a preceding interview series (for a counterfeit polo shirt, the associations were "design like the original product, quality very similar or hard to distinguish from genuine article, no warranty, priced EUR 10"; the generic product was characterized as "nice design, but less exclusive than branded product, good quality, with warranty, priced EUR 32"; and the original product was characterized as "exclusive, characteristic design, high-quality, with warranty, priced EUR 70"). The option to purchase no product in the second round was provided to sort out those respondents who were not familiar with the test model requirement to spend a given amount of money on a given product category. The test was repeated in a second round for an exclusive watch, a less exclusive genuine product, and an imitation of the genuine article.

Without budget constraints, 13% of the respondents chose the imitation product, 22% a generic product, and 65% the genuine, branded article. Repeating the test under the assumption that no counterfeit article was available, 32% chose the generic product and 68% the original (c.f. Table 7.1). With the consumers who chose licit products in the first round retaining their choice, 79% and 21% of the counterfeit consumers selected generic and genuine articles, respectively, when no counterfeits were available. Increasing the price of the counterfeit article reduced the willingness to purchase it, but increased the probability that those who had purchased it despite its higher price would have selected the genuine product if no counterfeit had been available.[37] Repeating the test with more expensive genuine articles (i.e. a Rolex watch for EUR 2,500, a counterfeit selling for EUR 30, and a watch of a less exclusive brand priced at EUR 300) showed a significantly smaller movement from counterfeit to genuine purchases.

---

[36] The questions were part of the survey described in Section 3.1.

[37] This statement stems from two tests among 60 participants where (1) the price of a counterfeit shirt was set to EUR 20 while maintaining the original's price, and (2) where the price of the counterfeit was left unchanged and the original's price was reduced to EUR 40. A survey involving more participants would be needed to increase the validity of the findings. Furthermore, the influence of the generic product's price was not investigated.

Table 7.1: Consumer decisions in markets with and without counterfeit goods

Consumer decision	Counterfeit article	Generic product	Genuine product
Counterfeit is available	13,0%	22,2%	64,9%
Counterfeit is not available	-	32,4%	67,6%
Share of counterfeit buyers selecting alternative (*)		79,2%	20,8%

*) Under the assumption that buyers of generic and genuine products do not alter their purchasing decision when counterfeit articles are no longer available.

The findings provide evidence of the existence of counterfeit consumers who buy genuine, branded articles if no corresponding imitation products are available. Furthermore, the probability of such consumers purchasing a genuine product appears to be negatively correlated with the relative price difference between the counterfeit and the genuine product.

Estimating the substitution factor among counterfeit and genuine articles

The model presented below explicitly focuses on the short-term impact of counterfeit trade on revenue, assuming that individual purchases – or purchases within a short period of time – do not affect future demand. The effects on future purchasing decisions are subsumed under a long-term analysis, which will be addressed in the following section on brand-related aspects. The separation of short and long-term analyses reduces the complexity of the design and aims to increase the accuracy of the results.

The substitution factor for the non-deceptive counterfeit case

Prices paid in non-deceptive counterfeit cases vary considerably depending on the quality and the risk associated with the product. Both the price of a genuine article and the corresponding counterfeit strongly influence the likelihood that a potential counterfeit consumer will actually purchase the original product instead of the imitation. In fact, the survey outlined above indicated that the probability of a counterfeit consumer purchasing a genuine article when no imitation is available increases with the ratio *price of counterfeit/price of genuine article*. This relationship helps to more precisely estimate the magnitude of the substitution effect. The amount of money non-deceptive counterfeit consumers spend on illicit products of a certain brand and type can serve as an approximation of the average amount they would have paid for the genuine good if no illicit products had been available (a phenomenon which may be referred to as constant budget assumption). As an example of the constant-budget assumption, of every ten consumers spending EUR 35 for a counterfeit handbag, one would buy the genuine accessory for EUR 350. The simple mathematical formulation is given in Equation 7.1, where $P_{Counterfeit}$ and $P_{Genuine}$ denote the cost of the counterfeit and genuine goods respectively, and S denotes the substitution factor.

$$S = P_{Counterfeit} / P_{Genuine} \qquad\qquad (7.1)$$

Figure 7.1 illustrates S for the constant budget assumption. The implications seem to be valid for counterfeits which, without closer inspection, may pass as genuine.

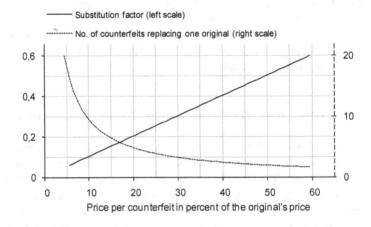

Figure 7.1: The substitution factor and number of non-deceptive counterfeits displacing one genuine product

The substitution factor for the deceptive counterfeit case

The substitution factor for deceptive counterfeiting is typically much higher compared to non-deceptive cases. Reasons are that the decision to buy the genuine product of a given type and brand has already been made, and that the price of the deceptive counterfeit product is often only insignificantly below the price of the original article.[38] With respect to the deceptive counterfeit case, consumers often believe they are purchasing genuine products that are on discount. Right holders and

---

[38] The latter is the case since low prices are likely to reveal the illicit status of the product and thus are avoided in the case of deceptive counterfeit goods.

manufacturers of licit products may have considerable knowledge of the price elasticity of their products, and therefore may develop more precise models based on their experiences. However, the constant budget assumption seems to be a good approximation of the present case as well. Please note that the substitution factor S is significantly higher compared to the non-deceptive case since the relative price difference between licit and deceptive illicit products is mostly small (c.f. Figure 7.2).

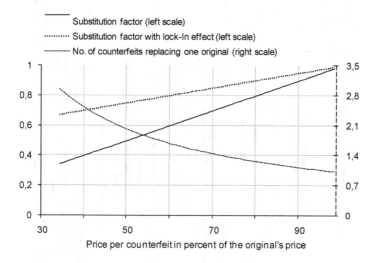

Figure 7.2: The substitution factor and number of deceptive counterfeits displacing one genuine product

Though the sample size of the survey and the variety of the products under study is too small to generalize the findings, the constant budget assumption seems to be valid for a wide range of deceptive and non-deceptive counterfeit cases. When estimating S, however, several additional product-specific characteristics may have to be considered. Strong lock-in effects or monopoly situations of essential goods can lead to considerably higher substitution factors, at least in markets where consumers have a high buying power. Nevertheless, the approximation appears to be superior to the established models which define S to equal either zero or one.

Computational framework

A model tailored to support practitioners in their efforts to derive loss estimates was developed based on the previous considerations. It was validated and instantiated in several interviews with industry experts. The computational rules are provided in Figure 7.3. The entry fields in Part A and Part B allow for a differentiation between three price categories among counterfeit goods, whose shares are calculated

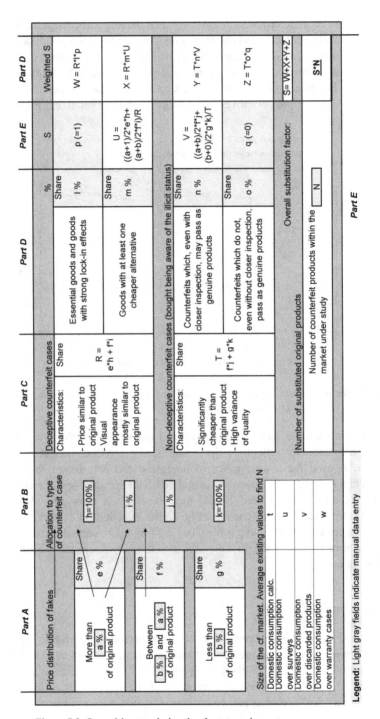

Figure 7.3: Spreadsheet to derive the short-term impact on revenue

according to the equations provided in Part C. Part D specifies market and counterfeit-related characteristics, for example lock-in effects, genuine alternatives and the visual quality of the imitation products. Part E defines the substitution factors for each subcategory which is weighted according to its relative share in Part D; the sum equals the overall substitution factor S. The product of S and the number of counterfeit articles N equals the amount of genuine products which are displaced by their illicit counterparts.

7.2   *A model to assess the impact on brand value*

Brands are an efficient means to efficiently reach large numbers of people, pledging that the associated good delivers a clearly stated, specific set of characteristics. Serving as a distinctive symbol for products, services, and organizations, brands not only constitute a reference to functional characteristics and physical properties, but can also convey values, feelings, expectations, and establish a relationship of trust between manufacturers and consumers. From a consumer's perspective, these references may constitute a considerable benefit as they help to reduce search costs, lower the perceived purchasing risk and may offer a means to express a desired self-concept. Consequently strong brands enable companies to charge considerable price premiums, stimulate and stabilize demand, reduce the risk of new product or service introductions, and strengthen their own position among the distributors. Without doubt, brands today are one of the most valuable assets of numerous companies (c.f. Aaker 1996a).

Counterfeiting undermines the concept of branding. The illegitimate disaggregation of brand and product has the potential to profoundly damage the relationship between licit manufacturers and consumers. In fact, many executives whom we worked together with regarded the implications for brand equity as the most important threat imposed by counterfeit trade. And the effects are indeed multifaceted. Brands are complex constructs, and the presence of illicit imitation products influences different elements in different ways. A high quality alkaline battery manufacturer may suffer from negative quality associations, while a label of a fashion handbag may seem to be less exclusive or special. Consequently an impact analysis has to provide a complete picture of the various different implications. Conclusions that are based upon an investigation of singular effects only are very likely to be misleading. As an example, recent publications that highlighted the beneficial influences of counterfeiting on brand value are insightful since they stress the potentially positive effects of imitations on brand awareness. However, as they ignore the effects on perceived exclusiveness, originality, etc., the findings should not alone be used to derive the overall impact on brand value.

Behaviorally-oriented brand valuation models are helpful to structure comprehensive analyses. They ascribe the value of a brand to a number of determinants – for example brand awareness, perception of quality, brand associations, brand loyalty and other brand assets – which not only reflect functional properties such as a reduction in search costs, but also emotional or interpersonal values.[39] As counterfeiting affects these determinants in various ways, managers should evaluate the effects on each determinant individually rather than try to assess the entire impact at once, but combine the findings into one well-balanced result at the end. A differentiated analysis helps to balance the positive and negative effects, and allows for the development of more specific courses of action to safeguard the brand.

Impact on the individual components of brand value

We outline below the implications of counterfeit trade on brand awareness, perceived quality, brand associations, and brand loyalty. Marketing specialists will most likely recognize some well-known principles that guide their everyday work. We nevertheless recommend revisiting some of these basic ideas to highlight the most important implications for markets with imitation products.

Brand awareness relates to the likelihood that the brand name will come to mind when given a product category or the brand-related needs are to be fulfilled. It is an important factor in consumer decision-making as it significantly increases the chance for a product or service to enter the frame for a purchasing decision, establishes a feeling of familiarity, and serves as a root for further brand associations. For certain product categories counterfeiting has the potential to significantly increase brand awareness. Fashion goods which use a trademark as a design element constitute a prominent example. Whether this ultimately leads to an increase in brand value is highly questionable and only likely in very special cases. Marketing efforts usually try to provide a well-defined, coherent picture of a brand, aiming at certain consumer groups. Counterfeit products are likely to destroy this image as companies cannot control where the products are sold, who is displaying or wearing them, which properties (for example in terms of visual quality) the product has, and which values the overall setting conveys. If the brand image generated by a counterfeit culture is not in line with the company's strategy, increased brand awareness is unlikely to translate into a higher brand value. If, however, the group of counterfeit consumers serves as some sort of role model for licit customers, imitation products may contribute to a brand's value.

The perceived quality constitutes an important factor in consumers' purchasing decisions. However, potential buyers often lack the ability or motivation to assess the quality or performance of a product. They frequently rely on secondary information and related associations to support their decisions. In this context brands can convey an expectation of durability, precision, grade and performance, or, in short, can serve as a sign or "seal" of the desired product characteristics. Brands

---

[39] C.f. Zimmermann et al. (2001).

can help to establish a relationship of trust between the manufacturer and the customer, which is strengthened if the products meet the customer's expectations, or gets damaged if they do not. For the consumer, the brand is a sign of quality and thus generates utility by reducing search costs and lowering the perceived purchasing risk. Introducing counterfeit products into a market subverts this key function of branding as consumers who purchase substandard products that bear the original's brand name can misattribute the low quality to the licit manufacturer. Companies know about the impact of low quality on future sales and invest a lot in quality management to gain a high level of consumer satisfaction. These measures may, however, prove ineffective if the original products are intermingled with a considerable quantity of deceptive counterfeits. From a consumer's perspective, non-deceptive counterfeits and substandard original articles are just the same. When assessing the impact of deceptive counterfeits on brand value, brand owners may apply the same reasoning as for genuine products which do not meet the required quality standards.

Brand associations not only refer to product-related characteristics, but also to non-product-related attributes. Brands can serve to express social status, wealth, individual taste, attitude, and distinction or membership in a certain group. They may therefore constitute a significant non-material benefit for the consumer. According to Vigneron and Johnson (1999) these benefits can be ascribed to different characteristics of human behavior such as: ostentation (Perceived Conspicuous Value), conformity (Perceived Social Value), non-conformity (Perceived Unique Value) and self-actualization (Perceived Emotional Value). Counterfeit trade affects these perceived benefits in various ways.

Perceived conspicuous value. Perceived conspicuous consumption serves to signal wealth, success, power, and status (Veblen 1899). It has an important influence on the development of preferences for many quality and luxury products, especially on those which are or can be purchased, consumed or used in a public environment. The perceived exclusiveness and the impression that only a small fraction of users can afford these goods is an important aspect of their associated value (Braun and Wickklund 1989). Consequently the price of products was found to have a positive effect on the perceived conspicuous value. Vigneron and Johnson (1999) state that, "Veblenian consumers attach greater importance to price as an indicator of prestige, because their primary objective is to impress others". If counterfeit goods cannot be distinguished from original products by visual inspection and are available for a fraction of the genuine product's price, they become attractive to those who are not willing or able to purchase the original product for the higher price. Illicit goods can provide a similar conspicuous, interpersonal value as long as the risk of being "convicted as a con man" is small, and are thus frequently purchased as non-deceptive counterfeits. At the same time the higher market share of the branded exclusive goods and especially the ownership by people who are not regarded as members of an "adequate" social group can reduce (and are likely to reduce) the perceived conspicuous value, and consequently reduce the price premium a licit consumer is willing to pay for the genuine article.

Perceived social value. The perceived social value, which influences the lower-end of brand extension, was referred to as the bandwagon effect by Leibenstein (1950). Similar to the other interpersonal values of brands, a desired group affiliation is the motivation for purchasing branded goods. However, here the conformity with the social reference group is the major characteristic of the effect. Leibenstein also covers cases where branded goods are explicitly not bought in order to distinguish oneself from another social group. Product prices seem to play a less important role within the conformity context. According to Vigneron and Johnson (1999), "relative to snob consumers, bandwagon consumers attach less importance to price as an indicator of prestige, but will put a greater emphasis on the effect they make on others while consuming prestige brands." Analyses conducted in the United States suggest that the number of consumers and the size of a group increase the perceived social value more than the culture or taste of a higher social class. The existence of counterfeit goods can dilute the perceived social value of branded goods. However, according to the concept of conformity, the impact of counterfeit products does not seem to be as severe as for the Veblenian effect or perceived unique value, which is discussed in the following paragraph. In certain cases, the existence of counterfeiting may even foster the perceived social value due to increased group size.

Perceived unique value. The perceived unique value of branded goods is often referred to as the "snob effect". The snob effect may occur during two circumstances: (1) when a new prestige product is launched, the snob will adopt the product first to take advantage of the limited number of consumers, and (2) when the majority of a relevant group cannot afford the product. The "snob effect is in evidence when status-sensitive consumers come to reject a particular product as and when it is seen to be consumed by the general mass of people".[40] Here, the limited numbers of consumers who are able to purchase or know of the existence of a brand constitute an important factor building the perceived unique value. Verhallen and Henry (1994) state that scarcity of products has a greater effect on demand when people regard the product as unique, popular and expensive, and Vigneron and Johnson (1999) show that "snob consumers perceive price as an indicator of exclusivity, and avoid using popular brands to experiment with inner-directed consumption". Similar to the conspicuous consumption of branded goods, product counterfeiting has a negative impact on the perceived unique value as it allows less-respected consumer groups to purchase branded goods and thus reduces their perceived exclusiveness and expensiveness. Moreover, the "snob" consumer group often has high claims regarding the quality of the product. Deceptive counterfeit cases with the resulting high probability of dissatisfied customers may therefore severely damage the value of the brand.

---

[40] Vigneron and Johnson (1990) following Mason (1981).

Perceived emotional value. Certain goods have a personal emotional value. Luxury products, for example, can provide subjective intangible benefits which are not necessarily related to communicating values or status to others. These goods are also bought for one's pleasure, for aesthetic reasons, excitement, or simply for enjoyment. The literature often refers to the hedonic effect when consumers value the perceived utility acquired from a brand that is caused by feelings and "inner satisfaction". According to Vigneron and Johnson (1999), "hedonist consumers are more interested in their own thoughts and feelings, thus they will place less emphasis on price as an indicator of prestige". The perceived emotional value is a result of the unity of brand and product. Purchasing of goods due to their perceived emotional value does not seem to be heavily affected by the decisions of other consumers. Non-deceptive counterfeiting is therefore unlikely to reduce the brand value in this context. However, hedonic consumers are likely to have high expectations regarding the quality of the product. Deceptive counterfeit cases may therefore again severely damage the value of the brand.

Brand loyalty – the biased behavioral response (i.e. purchasing decision) with respect to one or more alternative brands out of a set of such brands – is frequently regarded as the strongest measure of a brand's value. It is shown, for example, in high price premiums and repeat purchasing. As a complex construct in itself, brand loyalty is highly interrelated with the consumer's brand associations, quality perception, and other brand assets which are all influenced by counterfeit trade. However, counterfeiting has the potential to diminish the concept of brand loyalty itself. Since counterfeiting can be perceived as a disaggregation of brand and product, brand-prone consumers may purchase counterfeits which bear the brand name, but only mimic genuine goods. As buyers may purchase such articles knowingly, brand loyalty does not necessarily lead to increased sales for the licit manufacturer. When using brand loyalty as a key factor of brand value, one may therefore have to investigate the linkage between brand and product to validate the results.

Input from our consumer survey

An evaluation of the perceived impact of counterfeit trade on brand value was part of the consumer survey we introduced in Section 3.1. Within this study the participants were asked to evaluate the general implications of counterfeit trade on brand value from their perspective and to say whether they perceived the existence of counterfeit goods as a devaluating or annoying phenomenon. The distinction was made to arrive at the personal implications for individual respondents rather than to receive an answer which also reflects the anticipated effects with respect to others. In fact, about one half of the respondents thought that counterfeit goods reduce the value of genuine articles, whereas 40% felt that branded goods become less valuable for them if counterfeits are sold. 30% said they would be annoyed if cheap imitations of products they had purchased for a much higher price became available. The findings are illustrated in Figure 7.4, where the respondents are divided into two groups depending on their self-assessment of the frequency of purchases of genuine exclusive products.[41] The analyses of the equality of group means show no statistically significant differences between brand-prone and non-brand-prone consumers (c.f. Table 7.2).

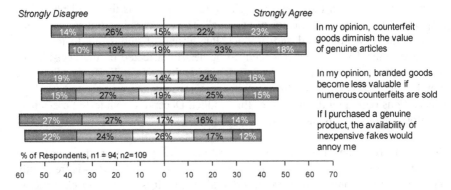

Figure 7.4: The perceived impact of counterfeiting on brand value; brand-prone consumers (top) vs. non-brand-prone consumers (bottom)

Table 7.2: The perceived impact of counterfeiting on brand value; analysis of the equality of group means

Independent variable	No intent. purchase		Intentional purchase		F-value	Signifi.
	mean	stdv.	mean	stdv.		
In my opinion, counterfeit goods diminish the value of genuine articles	0,16	1,40	0,30	1,26	0,59	0,444
For me, branded goods become less valuable if numerous counterfeits are sold	-0,09	1,39	-0,02	1,31	0,12	0,725
If I purchased a genuine product, the availability of inexpensive fakes annoys me	-0,36	1,39	-0,28	1,30	0,21	0,648

Computational model

The missing theoretical foundation of the proposed model – and, to a certain degree, also of the established brand assessment tools – may disillusion readers who are not actively involved in brand equity research. However, the frequent application of these tools is justified by their usefulness rather than by their theoretical foundation

---

[41] Brand-prone consumers agreed or strongly agreed with the statement that they frequently purchased genuine exclusive products, whereas non-brand-prone consumers disagreed or strongly disagreed. Respondents who took an intermediate position were excluded from the analysis.

(c.f. Aaker 1996b). Though different assessment tools may provide monetary values which differ by a factor of two or more for the same brand within the same market, they nevertheless prove helpful when comparing brands, when monitoring their development, or when estimating prices for selling or buying brand names. This is seen as a motivation for proposing a measurement tool which captures the impact of counterfeit trade. It is regarded as a first conceptual step, and users are encouraged to extend and adjust the concept according to their product and brand-specific requirements.

The following approach does not aim to replace established brand assessment tools, but is designed as an add-on for established business-finance-oriented or composite business financial/behavioral models. It constitutes an additional computational step after the overall brand value has been determined to separately show the impact of counterfeit trade.

Existing brand equity models, if designed carefully, take into account loyalty measures such as price premium and customer satisfaction, perceived quality measures, factors regarding association and differentiation (such as perceived value and brand personality), as well as brand awareness and market share. Since most modern tools focus on the customers' perception with respect to the overall value, the impact of counterfeit trade is implicitly captured – but not shown separately – by the calculations. Nevertheless, it is desirable to determine the extent to which the existence of counterfeit goods has reduced or possibly increased their overall value, or, in other words, how the brand value would be affected by a reduction in counterfeit trade.

The proposed extension is based on the personal and interpersonal values which the brand under study may exhibit for consumers, and takes the share of deceptive and non-deceptive counterfeit cases as input variables. The overall calculation process is divided into three steps:

- First, the brand value is calculated using a conventional model which implicitly includes the impact of counterfeit trade.
- Second, the impact of counterfeit trade is derived with the proposed model; the following paragraph details the exact procedure.
- Third, the fictional brand value without counterfeit trade is obtained by dividing the brand value derived in step one by the divisor calculated in step two.

It is now possible to calculate the difference between the brand value with and without counterfeit trade through simple subtraction. The complete model, with a focus on step two, is illustrated in Figure 7.5.

The perceived values of brands from a consumer's perspective constitute the starting point for the calculation of the impact of counterfeit trade. Equation 7.2 shows the computation rule.

$$ICT = 1/(1 - \Delta S_d \cdot Q + \Delta S_{nd} \cdot (-V \cdot \tilde{v} - S \cdot \tilde{s} + B \cdot \tilde{b})) \qquad (7.2)$$

Info Box 7.2: Estimating the model's parameters

Our brand impact model requires a set of parameters (for example the percentage of consumers who are primarily driven by Veblenian, Snob, and Bandwagon effects) as well as data on the overall share of deceptive and non-deceptive counterfeits as an input. We know that the data is difficult to obtain. However, with support from marketing, potential input from own surveys, and a good feeling for consumer choice with respect to own products, it is feasible to estimate the parameters with a sufficient level of accuracy. It may be necessary to adjust the parameters over time, but this is just the way such tools evolve.

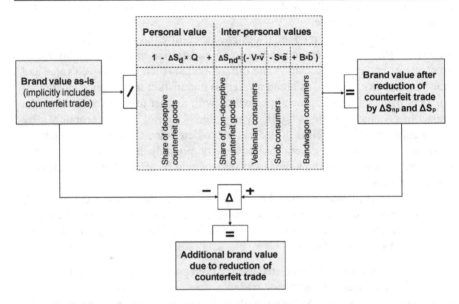

Figure 7.5: The brand evaluation tool to estimate the impact of counterfeit trade

ΔS_d refers to the share of deceptive counterfeit goods within a market, whereas ΔS_{nd} denotes the non-deceptive counterfeit cases. The parameter Q expresses the importance of meeting the customers' quality expectations. Including Q is necessary since consumers may be more forgiving to some brands than to others. Q can be estimated as the reciprocal value of the share of substandard goods that the majority of customers would accept before losing confidence in a brand. It may be as low as one percent for car parts ($Q = 50$) or may approach 50% for luxury accessories[42] ($Q = 2$).

---

[42] Please note that Q refers to deceptive counterfeit cases and thus to the way consumers who try to purchase genuine products respond to substandard quality.

Parameters V, S, and B denote the share of consumers who are primarily motivated by Veblenian, Snob, and Bandwagon effects respectively. Non-deceptive counterfeits have a negative impact on the perception of Veblenian and Snob consumers, and a positive impact on those attracted by the Bandwagon effect. The weight factors $\tilde{v}$, $\tilde{s}$, and $\tilde{b}$ can be used to adjust the model according to brand- and product-specific properties. As a starting point users may set them to 1.0. In a sample calculation, the approach provided realistic estimates for a fast-moving consumer goods company, as well as a manufacturer of the brand associated to an exclusive luxury product.

7.3   Implications for quality costs, liability claims, and future competition

According to customs officials and interviews with brand owners, hardly any counterfeit articles seized by customs within the time frame of this research would have satisfied the quality standards of the corresponding licit manufacturers. Not all counterfeits were dangerous or without function, but some of the goods – brake pads, pharmaceuticals, baby food, airplane spare parts, etc. – even imposed a severe risk to consumers' health and safety. For affected enterprises deceptive counterfeit cases ultimately lead to increasing numbers of warranty claims, product recalls, dissatisfied customers, or claims for indemnification. Their associated costs are discussed in the remainder of this section.

Impact on quality costs

Especially in manufacturing, companies are aware of the impact of substandard products on the overall costs and therefore invest considerable efforts to ensure high levels of quality and a low number of defects among their turnout. Quality management techniques such as six-sigma (Taguchi and Clausing 1990) or Poka-Yoke (Shigeo 1986), which strive for extremely low error probabilities, all follow the line of argument that high average follow-up costs of defects justify higher spending on quality assurance.

Since it is the buyer or consumer who ultimately judges the quality of an article, one may argue that quality management should not end at the manufacturer's dock door, but rather include measures to ensure the authenticity of a product at the customer's end. However, if original articles are intermingled with imitation products, established quality measures may prove ineffective. Especially when being sold through trustworthy channels, counterfeits are often mistakenly regarded as genuine articles and built into otherwise genuine products, with potentially detrimental effects on the reputation of a company and high costs to limit the resulting damages. From a customer's perspective, deceptive fakes and substandard original articles appear to be the same. When assessing the impact of deceptive counterfeits, it is therefore legitimate to apply similar standards as to defective genuine products.

In their most basic form total quality costs are expressed as the sum of the costs of avoiding defects (prevention), finding defects by inspection, audit, calibration, test and measurement (appraisal), and the consequences of actual defects (failure) (c.f. Montgomery 2004). Failure and appraisal costs typically have a negative slope and a positive second-order derivative with respect to the degree of perfection, where the latter is defined as one minus the failure rate. Expenses for the prevention of failures have a positive slope and are growing increasingly fast with the degree of perfection approaching 100%. Most explanatory models assume that a minimum of the total quality cost function exists for a specific failure rate.

The financial impact of counterfeit articles with respect to the total cost of quality is illustrated in Figure 7.6. The introduction of counterfeit articles into licit supply leads to a decreasing level of perfection (1) → (2). Maintaining a constant level of preventive measures, indicated as a shift in the corresponding curve (3) → (4), results in an increase in failure and appraisal costs (5) → (6), and therefore to an increase in the total quality costs (7) → (8). As the graph illustrates, the new total quality costs can be reduced by increasing the spending for preventive measures (i.e. on supply chain security) up to the point where the total costs reach their new minimum, with (4) → (9) and (8) → (10), respectively.

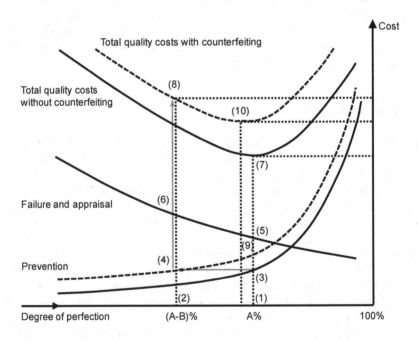

Figure 7.6: Quality costs and the impact of counterfeit trade

Supported by company- and product-specific data, the model allows for a more analytical approach in its attempt to determine the optimal spending for anti-counterfeiting measures. Moreover, it provides some arguments for including anti-counterfeiting efforts in the quality management domain.

Liability claims

As outlined before, counterfeit articles can lead to threats to the health and safety of users or consumers. Claims for indemnification may result from customer assertions that the brand owner or licit manufacturer did not take sufficient preventive measures, or may be based on the false assumption that a counterfeit was a genuine article. Once used or consumed, it is difficult to prove the illicit nature of a product. In the case of counterfeits infiltrating the licit supply chain and thus being sold through trustworthy channels, the burden of proof for companies is even more severe.

In general the financial impact due to liability claims is difficult to estimate in advance. How a company fares in the face of such threats depends on the type of product and the possible damage fakes can cause. In any case the impact of counterfeits should be on the risk radar of companies. Moreover, the anti-counterfeiting efforts must not be below industry-standards and should be well-documented.

Impact on future competition

A second-order effect, the emergence of counterfeiters as licit competitors, is starting to attract a great deal of attention. Counterfeit production enables illicit actors to build up know-how in manufacturing and distribution. Most counterfeit automotive spare parts which are manufactured in China, for example, are not bound for export to western countries, but serve the Chinese market. With a share of approximately 30% of selected brands,[43] one cannot expect the illicit factories to disappear once intellectual property rights are more strictly enforced. They are more likely to turn into licit providers serving the aftermarket or even to become full-fledged manufacturers in the future.

The Chevrolet Spark provides a good example of China's product-copying talent. Less than one year after the Spark was introduced, Shanghai Automotive Industry Corp released its Chery QQ, which was strikingly similar to the former. During a visit to Shanghai, U.S. Congressman Sensenbrenner commented on it: "It's such a knockoff that you can pull a door off of the Chevy Spark and it fits on the QQ – and it fits so well that the seals on the door hold." Another example is Allied Pacific Motor's Comel Manja JMP 125, an exact copy of Honda's CG125

---

[43] The estimation is based on data provided by the International Chamber of Commerce (ICC 2007).

motorcycle. The knock-off severely damaged sales of the original manufacturer in the Asian market, while Honda's lawyers stood rather helplessly aside. The High Court of Malaysia has recently dismissed the application by Honda for an interlocutory injunction against Allied Pacific Motor. As Ling (2005), a Malaysian IP expert, puts it, "the case (...) is a solemn reminder to IP owners that even in a factually strong case of infringement involving replicas or imitation of products, the court may not necessarily be prepared to grant interim relief pending the full trial of the matter. (...) (S)ocial justice was held in the eyes of the court to prevail over the purely commercial interest of the plaintiff. The resounding message that emanates from the case is that the larger the operations of the IP infringers, the easier it is to tilt the balance in defeating an interim injunction application."

Intellectual property theft as a means to accelerate learning processes and foster demand is neither a new phenomenon in the corporate world nor a new field of re-search (for example Mansfield et al. 1981 and Helpman 1993). Companies participating in the first outsourcing wave to Asia in the 1980s frequently reported patent and design infringements which even influenced decisions on further relocations abroad. However, intellectual property theft in the 1980s mainly affected companies which were engaged in outsourcing activities, i.e. which actively transferred know-how to partner companies in foreign countries. Counterfeit production today concerns a wider range of enterprises. Due to high re-engineering capabilities and easier access to modern production facilities, illicit actors today no longer rely on internal-company information to produce counterfeit goods. As a consequence, brand owners and manufacturers should also consider taking active steps against illicit actors from a competitive strategy point of view, rather than just staying out of the market.

7.4   Positive effects of counterfeit trade

Counterfeit trade is not always bad for the manufacturer or brand owner. In fact, several positive effects may occur, mostly resulting from a higher perceived market share and an enhanced accessibility in lower-price segments. These implications can be ascribed to the following categories: positive brand-related effects, network effects, and lock-in effects.

Positive brand-related effects may exist as the presence of counterfeit goods can foster brand recognition and awareness. However, they are mostly canceled out by negative implications; for example on the associated quality and perceived exclusiveness. A detailed discussion on the overall implications for brand value has been provided in Section 7.2.

Network effects cause a good or service to have a value to a user or consumer which is positively related to the number of users already owning that good or using that service. Examples of applications with strong network effects are fax or email systems, where the purchase or participation by one individual indirectly benefits other users. This type of side effect in a transaction is known as an externality. Externalities resulting from network effects are referred to as network externalities. In the case of modern communication devices, distinct network externalities which are particularly strong if the cost of additional entities (for example due to the communication overhead) is low can be frequently observed (c.f. Fleisch 2001). However, similar effects also exist in other domains. The Bandwagon effect refers to the observation that people often tend to do things because others do the same, and people's preference for a commodity may increase as the number of people buying it increases (Leibenstein 1950) (again, c.f. Section 7.2. for a discussion of the Bandwagon, Veblen, and Snob effects). If counterfeit articles are compatible with their genuine counterparts and are used within the same domain, i.e. by corresponding users or systems, they can foster network externalities similar to additional genuine articles.

Lock-in effects refer to customer retention over switching costs and a systematic market foreclosure against competitors. Customer lock-in effects often result from the perceived or real effort of changing to another product or service. Expected switching costs may result from searching for and trying new products, set-up costs, and the effort to learn how to use the new product. Companies often succeed in establishing lock-in effects when selling durable components which require complementary products (for example electric toothbrushes and replacement brushes, computer systems and platform-dependent software, or printer and toner cartridges). Market foreclosures against competitors often build upon strong customer lock-in

---

Info Box 7.3: Lock-in effects in the Chinese software market

About 90% of all programs in the Chinese software market are not legitimately licensed (BBC 2005). The vast majority of personal computers use Microsoft Windows as an operating system, which, as a genuine product, is sold for a multiple of a Chinese white-collar worker's monthly average income. Needless to say, if no illicit copies were available, only a fraction of today's PC users in China would be familiar with Microsoft's product and would rather use open source software such as Red Flag Linux. Now, after the Chinese government required computers manufactured within the country's borders to have pre-installed authorized operating software systems when they leave the factory, Microsoft can build upon a large user base and use its strong market position to generate revenue. The stakes are huge as China has become the world's second-largest PC market, with more than 19 million PC shipments in 2005 (Gartner 2006). In an interview with CNN, Bill Gates stressed the beneficial effects of software piracy on the development of Microsoft's market in China, mainly due to lock-in and barriers to entry for emerging legitimate competitors (Kirkpatrick 2007).

effects and are also frequently related to network effects. In economies with low household incomes, counterfeit articles can constitute an efficient way to familiarize large numbers of users with goods which they could not afford as genuine products at that point in time. Counterfeits can help to create barriers to entry for potential competitors as knockoffs of mostly high-priced genuine articles compete against lower-end and often domestic competitor products. When a country starts to enforce its intellectual property rights more consistently, which is to be expected during its economic development alongside increased consumer buying power, the brand owner is able to leverage its prevalence in the market (see Info Box 7.3 for an example).

7.5   Research on the impact of counterfeit trade

In the academic community, very little has been published investigating the financial impact of counterfeit trade on affected companies. Beyond that, the findings are in part contradictory. Nia and Zaichkowsky (2000) surveyed the perceived value of luxury goods where faked counterparts are available in the market. They found that the majority of respondents indicated that the value, satisfaction, and status of original luxury brand names were not decreased by the wide availability of counterfeits. Furthermore, the majority of respondents disagreed that the availability of counterfeits negatively affects their purchasing intentions for original luxury brands. Feinberg and Rousslang (1990) conducted a study among U.S. companies, examining the welfare effects of foreign intellectual property rights infringements. While they do not specifically focus on counterfeit trade, they find that profit losses are at least as great as 1% of the total sales and expenditures on countermeasures are less than 4% of the losses.

Barnett (2005) and Yao (2005b) highlight that companies may experience increased brand awareness as well as additional demand due to bandwagon and network effects. Learning effects among illicit companies were illustrated in a German Newspaper (Die Zeit 2006), but have not yet been thoroughly investigated by academics.

In fact, compared to the importance of the phenomenon, research on the economic impact of counterfeit trade at a microeconomic level is sparse. The required analyses are in part strongly related to investigations of the counterfeit demand-side, as alternative buying behavior and the perceived impact on brand value have to be first understood before conclusions can be drawn on the potential financial losses amongst brand owners and licit manufacturers.

Table 7.3: Research on the impact of counterfeit trade

Author(s)	Year	Short description
Harvey/ Ronkainen	1985	– Discussion of potential ways illicit actors can obtain classified information which enables them to produce counterfeit articles. – Loss estimates based on industry estimates.
Liebowitz	1985	– Beneficial effects of intellectual property rights violations that result from the use of modern copying technologies. – Case example of the impact of photocopying on journal publishers.
Grossman/ Shapiro	1988a	– Demand-price curve in markets with both deceptive counterfeit articles and genuine products. – Welfare analysis regarding the disposition of confiscated counterfeit goods.
Grossman/ Shapiro	1988b	– Description of non-deceptive counterfeiting as a disaggregation of brand and product. – Demand-price curves in a market with counterfeit and genuine products.
Feinberg/ Rousslang	1990	– Welfare implications regarding foreign infringements of intellectual property rights.
Conner/ Rumelt	1991	– Analysis of not protecting software against piracy, which can be the best policy as it may raise profits and lower selling prices and is beneficial for both firms and consumers.
McDonald/ Roberts	1994	– Transfer of technology to less developed countries and the satisfaction of market need as a positive aspect of counterfeit trade.
Givon et al.	1995	– Diffusion modeling approach to track diffusion of pirated and legal software over time.
Wilke/ Zaichkowsky	1999	– Discussion of the impact of brand imitation on innovation and brand equity.
Nia/ Zaichkowsky	2000	– Perceptions of luxury brand owners towards counterfeit luxury goods.
Barnett	2005	– Discussion of the effects of counterfeiting on the perception of status goods.
Yao	2005a	– Impact of adopting a monitoring system on counterfeiting at a macroeconomic level by providing an economic model to assess the socioeconomic welfare effects due to counterfeit trade.
Yao	2005b	– Impact of the presence of counterfeit products on monopolist profits.
De Castro et al.	2006	– Benefits of product piracy on the right holder due to a reduction of inimitability, which can increase the overall value of the firm.
Montoro-Pons/ Cuadrado-Garcia	2006	– Substitution effects of piracy on legal demand for music recordings.

Part D   Product-protection Technologies

8 Principles of Product Security Features

Technological security features constitute an integral part of many anti-counterfeiting strategies. If properly deployed, they can strengthen the security of supply chains, hamper production and distribution of counterfeit goods, and help to prevent consumption of illicit articles. They primarily serve as a means to

- authenticate genuine goods, thus helping supply chain partners, customs, or users to distinguish genuine goods from counterfeit articles and
- increase the production costs among illicit actors, who may have to invest considerable effort to deal with security measures, thus making the protected product a less attractive target.

Many of the established authentication features are extremely secure in a sense that producing a perfect copy is (almost) impossible.[44] However, the benchmarking study presented in Chapter 4 clearly expressed the lack of confidence in the established security features among many anti-counterfeiting specialists. In fact, even the most advanced techniques have not been able to stop the recent growth in counterfeit trade – for a number of simple but often overlooked reasons. In this chapter we describe why these techniques often prove ineffective and what brand owners can do to overcome the shortcomings. We start with a short classification of the established security features, explain how counterfeit producers respond to these challenges, and provide a requirement analysis to support brand owners in selecting effective security measures. The results may require brand owners to fundamentally revise their approach to protection technologies. However, we are convinced that the underlying principles will be the foundation of future supply chain security measures.

8.1   Classification of protection technologies

Product security features can be broadly divided into disabling and identification technologies. Disabling technologies[45] aim to directly prevent illicit actors from reproducing the protected product or to confine the functionality of faked articles. Copy protection schemes for DVDs or activation processes for registered software

---

[44] Though the slogan "if you can make it, they can fake it" is probably true it is, in this context, somehow misleading. With sufficient means, it may sooner or later be possible to trick most approaches. However, for carefully designed techniques, the benefits from cloning such features (i.e. the earnings from counterfeit trade) will almost certainly be not worth the effort. As we will show later, an insufficient level of security is not the reason for the failure of many approaches.

[45] Or enabling technologies, depending on the perspective.

are prominent examples. A drawback of such measures is that they can only be applied to a small subset of products. It may be possible to disable imitations (or better: enable only genuine versions) of electronic devices or digital media, but the vast majority of goods cannot be "activated" or "deactivated" – just think of handbags, beverages, cigarettes, etc. If disabling features can be built in at all, their design very much depends on the product they have to protect (i.e. on the way its utility can be eliminated), and therefore have to be taken into consideration during an early phase of product development. Another drawback is that they tend to prove ineffective in the long run. Counterfeit producers may not be able to exactly rebuild the protection mechanism, but often succeed in developing a satisfactory workaround. The weakest link is rarely a feature's level of security but the insufficient (possible) depth of product integration.

Far more common are identification technologies.[46] They facilitate the recognition (or more precisely the authentication, i.e. the check of accuracy of the stated origin) of genuine articles and, conversely, the detection of imitation products. Identification technologies may be grouped according to their conspicity/accessibility, the depth of integration, level of integration, and the intrusiveness of individual tests.

Conspicuity/accessibility. Security features may be overt (they can be seen by others) or covert (they are hidden and only insiders should know how to find and use them). There has been dispute for quite a while now as to which approach should be preferred. Proponents of overt technologies say that their approach is more secure as the underlying principles can be made public and thus be reviewed by many experts. In computer and network security their view seems by now to be widely accepted. The advocates of covert features say, however, that concealing the feature and its principles adds an extra layer of security to the system as illicit actors may not even know what measure to break. While the latter position sounds plausible, its realization comes with one major drawback: security features should be frequently inspected, and inspections should take place at many stages of the supply chain. This requires that many people know about the covert feature – and is contrary to the idea of keeping security measures secret. Covert features may effectively complement a set of protection techniques that are applied to a product. However, they do not facilitate frequent and effortless authentication by a large number of users.

Depth of integration. Security features can be attached to a product's packaging, the product itself, or they may become an integral part of the object that is to be made secure. The approaches vary in cost (features can be attached in a cost-efficient manner to a product's packaging where they are also easily accessible during inspection) and the level of security (when tagging packaging it is in fact

---

[46] We also refer to them as authentication technologies when we want to stress that the genuine origin of an item is proven.

not the product that is authenticated, whereas features that are an integral part of the product are more difficult to tamper with). Examples of the three categories are holograms attached to packaging, micro-printings on a product and chemical markers in plastics molding material.

Level of integration. The level of integration denotes the number of objects that use identical security features. Manufacturers may use identical instances for the entire brand, or different instances for each product line, the production series, or for each individual item. The latter requires the features to contain elements that can be arranged to represent a large number of items. Holograms that include an imprinted number, sequential numbers printed in security inks, or specific Radio Frequency Identification (RFID) tags are examples of such technology. Item-level security has some important advantages over conventional approaches. If designed properly, breaking one instance of a feature does not compromise the entire system but enables the illicit actor to produce *only* many pieces of the one instance that has been analyzed. This is bad enough – but much easier to handle from the perspective of a brand owner who has to watch out for items that carry the compromised instance only.

Dynamic vs. static features. Test procedures are dynamic if they can easily accommodate changes in the inspection routines once the security feature is applied and the product is sold. Such changes are helpful when a feature has been compromised. With dynamic features at item level it becomes possible to block out individual identifiers remotely. The classification into dynamic and static is not common simply because there are not many established dynamic approaches. We will discuss their advantages in greater detail later.

Intrusiveness. Product inspections can be divided into destructive and non-destructive tests. Under destructive inspections we subsume scenarios where the product or its packaging is destroyed, i.e. where the article cannot be sold on a regular basis after the inspection has taken place. Examples are chemical analyses of drugs or tests for biological markers in fluids where sealed bottles have to be opened. Destructive tests tend to be expensive and are not suited for large-scale inspections (for example where companies aim to authenticate every fourth or third product).

Level of security and cost are, as a matter of course, two further important properties that characterize identification technologies. With level of security we refer to the effort that is necessary to reproduce (i.e. counterfeit) a security feature such that it will pass inspection. For advanced item-specific technologies one has to distinguish between cases where illicit actors succeed in reproducing only one feature and cases where they break the entire system. Costs include the cost per feature and its application to an item as well the expected costs per check. The latter are often overlooked but nevertheless important when companies really want their security mechanisms to be used.

There are numerous types of security techniques available on the market. Let us just look at one type that is frequently applied to a wide range of products, holograms. Holograms were once very difficult to produce and, with their strong visual appeal, served as an "eye catcher" on many products. They come in many different

---

Info Box 8.1: Product-protection technologies

Let us briefly describe four technical approaches that are frequently applied to authenticate products: optical features, chemical and biological markers, electronic features, and the use of object-specific characteristics.

Optical features such as holograms and security inks are based on reflection, refraction, diffraction, and absorption phenomena. Protection against imitations results from difficult-to-produce material or difficult-to-control manufacturing processes. The level of security, for example of holograms produced using dot matrix technology, can be very high. However, most features either require the inspector to have dedicated test equipment at hand or to know exactly how the genuine feature should look.

Chemical and biological markers are becoming increasingly attractive as an anti-counterfeiting measure, mostly due to the improved understanding of the unique characteristics of proteins, enzymes and DNA, and the ability to reliably detect traces of these additives. Advantages of such markers are that they can directly identify a product rather than only its packaging. Disadvantages are that tests are often destructive (bottles have to be opened or pills have to be taken out of their blister packs). Moreover, the tests require dedicated equipment and are rarely suited to large-scale inspections.

Electronic measures are based on cryptographic approaches where an electronic tangent can perform authentication routines or utilize a backend system that, with an item-specific object identifier, keeps track of the product's history. We will discuss electronic measures in greater detail in the following section.

Object-specific tests directly rely on unique, often item-specific characteristics of products. Individual surface structures, back-scattering characteristics of the used material when exposed to radio waves, etc., are recorded and associated to every instance of a product. The approach is comparable to taking electronic fingerprints or iris scans for identity checks. Changes in product design are rarely required. A (less charming) variation of the approach is to generate "finger prints" of security labels that are then attached to the item. Object-specific tests are non-destructive and can be highly secure. However, a drawback is the need for sophisticated and often product-specific test equipment.

sizes, shapes, levels of sophistication, etc. Holograms are overt, i.e. customers can (and should) recognize them, but may also contain covert properties (for example small structures that cannot be seen without additional equipment). They are mostly attached to a product's packaging but can sometimes also be affixed to the product itself. Inspections are non-destructive in a sense that they can be conducted without reducing the value of a product. Moreover, they are cheap to produce and easy to deploy. And they are rather secure. Reproducing a state-of-the-art security hologram such that it is indistinguishable from the original version is a challenge most illicit actors cannot overcome. So what is the reason why they do not effectively stop counterfeit trade? The cost (or effort) that is required to verify them thoroughly. As we will see in the next section, this drawback allows counterfeit producers to effectively deal with holograms – and with many otherwise secure features as well.

8.2   Attack scenarios and their implications

Most technologies are highly secure in a sense that it is extremely difficult for illicit actors to build exact copies of protected, genuine products. However, as we pointed out before, the technologies have not been able to stop the growth in counterfeit trade. While many brand owners used an ever-wider variety of ever-more sophisticated overt and covert features, illicit actors were able to rely on overwhelmed inspectors who could hardly cope with the multitude of brand- and product-protection techniques.[47] Even more than the exact duplication of security features or the reapplication of formerly used, genuine protection mechanisms, counterfeit producers follow a tag omission and obfuscation strategy to smuggle and sell their goods. A more formalized analysis of the different approaches is outlined below.

Feature/tag omission, i.e. the omission of the security features when imitating products that normally carry such features relies on low inspection rates and insufficient attention paid to protective measures. Feature omission is common. The phenomenon shows the need for large-scale – and consequently low-cost – inspections. Features that are applied but rarely inspected may exclude one's liability but do apparently not stop counterfeit producers. Inspections should preferably be automated even in loosely-guided processes for example in warehouses, at customs, and at retail stores.

---

[47] Needless to say, experts that are responsible for the anti-counterfeiting measures of their brand can be expected to know all of the applied features. However, other important stakeholders including customs and end users who have to deal with many other products cannot.

Obfuscation connotes the use of misleading protection technologies. In practice licit companies frequently change their security features to prevent counterfeiters from copying or cloning their protection technology. While following this paradigm of "creating a moving target", the licit parties unintentionally complicate the inspection process. Third parties in particular can be overwhelmed by the coexistence of different, mostly visual security features. Consequently counterfeit producers can often rely on the lack of knowledge (and the lack of time and motivation to acquire it) during inspection processes. It is rather common that counterfeit producers use security mechanisms which are not related to the genuine product. They may use some basic hologram instead of a highly secure version with concealed images or flip colors instead of micro printings. Who – besides the brand manager and his or her team – can tell the difference? The need to change anti-counterfeiting primitives when they become ineffective complicates the situation even further. Sometimes even genuine products from different production series carry different features. How could a customs officer with the huge amount of goods that has to be handled possibly deal with this situation? In fact, it is no surprise that the confidence in technological anti-counterfeiting solutions is rather limited.

In anti-counterfeiting systems that rely on more than one component, threats may not only originate in bogus product-security features, but also in malicious backend systems. When a barcode, a micro printing, or an RFID transponder references a database containing track-and-trace information or advanced shipment notices, the authenticity of the relevant source has to be verified.

Cloning refers to the precise duplication of security features so that the duplicates are almost certain to pass an inspection. In a system with cloned entities, investigators (or reading devices) can no longer ensure that the distinguishing mark they observe originates from the correct source; moreover, without taking the existence of duplicate features into account, observers would even falsely certify the authenticity of bogus components. Large-scale tag cloning attacks severely compromise the anti-counterfeiting solution. Fortunately, most security technologies offer a high level of protection and such incidents are not very common. However, the level of security still has to be carefully evaluated during the technology selection process.

Removal-reapplication attacks refer to the application of genuine security features from (mostly discarded) genuine products to counterfeit articles. This constitutes a potential threat for tagging technologies where security features are attached to an object (like holograms or RFID transponders) rather than made an inherent part of it (such as chemical markers). The consideration of this attack is of importance especially when protecting high-value goods like aviation spare parts, which often become accessible to illicit actors when they are discarded. When relying on tagging technologies, a defense is to tightly couple the security feature to the object, for example by tamper-proofing its physical package or by establishing a logical link between the object and the tag.

Denial-of-service attacks may be defined as "any event that diminishes or eliminates a network's capacity to perform its expected function" (Wood and Stankovic 2002). Since established anti-counterfeiting technologies usually do not yet rely on network resources, this attack is new to the brand- and product-protection domain. However, when authentication processes involve entities in disparate locations, the access to these resources may be disturbed. With respect to track-and-trace solutions (which we discuss later), attacks can cut off the connection between individual transponders and reading devices. When illicit actors target major distribution centers or customs, for example at harbors or airports, denial-of-service attacks may severely slow down inspection processes and thus interfere with the unobstructed flow of goods.

All of these attack scenarios should be considered in the design process. However, let us make it very clear that many product-protection technologies fail not due to their lack of security but due to their impractical inspection processes. The slogan "if we can make it, they can fake it" may be right, but something like "if we don't use it, they don't need to" seems to be more appropriate.

8.3 Requirements for security features

The review of common attack scenarios in Section 8.2 led to a set of extremely important requirements. They include:

- measures to avert a duplication of security features,
- a tight coupling of the security feature to the object that has to be protected,
- simple, standardized inspection routines that do not overburden inspectors, and
- efficient inspection processes at low cost even in loosely guided processes.

While the first two relate to traditional security aspects, the latter emphasize the need for a user-friendly design and reflect the limited resources during inspection. In addition to this set of exogenous requirements, a number of – partly interrelated – conditions are induced by the brand owners. They mainly relate to cost, product design, production processes, and potential future development to respond to more advanced counterfeit actors.

Different levels of security. The desired level of security has a major impact on the fixed and variable costs of the solution. It can be determined (1) by the risk or cost resulting from a compromised system, and (2) by the lifetime of the object which is to be protected. Risk or cost can be classified in terms of the potential health and safety hazards for consumers, or the incremental financial losses of licit manufacturers and brand owners. Depending on the probability of individual occurrence, health and safety hazards may require highly secure and expensive systems. If illicit products primarily cause incremental financial losses (for example due to dissatisfied consumers and substitution effects), a detailed cost-benefit analysis is helpful in order to select an appropriate protection mechanism. In many cases solutions with different levels of security and different costs will co-exist. This should not make us waive one of our core-requirements, the user-friendliness of the corresponding inspection routines. In an ideal setting, test procedures look the same regardless of the underlying security approach.

Manufacturing requirements. Existing manufacturing settings are often highly optimized with respect to throughput and down times. The addition of supplementary process steps can severely impact the key performance measures of the production facilities. This is especially the case in high-volume production environments for example in the pharmaceutical or fast-moving consumer goods industries. Here the required line speeds severely limit the choice of technology. Process steps that are necessary to integrate security features have to be as non-intrusive as possible.

Product-specific requirements. Product-related characteristics can impose a number of additional constraints on the choice of technology. Restrictions may result from the available size of security features, the object's material, and operating conditions such as temperature, electrical discharge, abrasion etc. When the security features are to be deployed at an early stage of the production process, aggravated conditions may apply (for example high temperature and pressure during injection molding, etc.). The product-specific requirements have to be analyzed on a case-by-case basis at an early stage of the design process.

Invariance of the product design. In order to enhance the level of security, it is desirable to integrate the security features in the product and not to rely on tagging its packaging. However, companies are rarely willing to subordinate product design to anti-counterfeiting measures. This limitation may further complicate the tag-in-product integration and often leads to selecting measures to authenticate the package rather than the product.

Migration path. Anti-counterfeiting technologies constitute a barrier for illicit actors only for a limited, unknown period of time. Consequently it is desirable to have the opportunity to change the underlying security primitive at low cost, i.e. without the need to alter the technical infrastructure or to require the user to get accustomed to new checking procedures. Again, the changes must not complicate the inspection process.

Requirements with respect to the inspection process. Individual security approaches may be chosen due to the specific advantages they exhibit. Techniques that allow for automated inspections, for example, may have to work at defined line speeds, distances to the reader, or maximum cost per inspection. Brand owners may also have to specify maximum acceptable failure rates, useful economic life, recycling requirements, etc. If special devices are required to perform the inspection, the future availability of such tools has to be assured.

Confidentiality. Last but not least, produkt-protection approaches shall not reveal confidential information of the manufacturer nor infringe the privacy of the user or consumer. Item-specific features contribute to the security of a solution but should not allow competing companies to draw conclusions on production output. More-over, proving the authenticity of a product without opening its packaging can be crucial to establish inspection processes at low cost. However, identifying an arti-cle in a consumer's shopping bag constitutes a threat to his or her privacy – and may result in a negative headline on the business principles of the brand owner.

The list of requirements is in fact comprehensive. In the following chapter we will describe how RFID technology can be used to develop an appropriate system.

9   The Potential of RFID for Brand- and Product-Protection

Radio Frequency Identification (RFID) is an automatic identification technology that relies on radio waves to transmit data, typically a serial number, between an RFID transponder (or tag) and a corresponding RFID reader. The technology is well established in applications such as animal tracking, vehicle immobilization, access control and payment systems. In recent years its potential to improve supply chain processes has generated considerable attention. Enterprises from diverse industries are hoping RFID will provide solutions for a wide range of management problems. Applications include approaches to increase processing efficiency for the receipt and dispatch of goods, improvements in process control and product quality, and savings due to faster and better information processing. The current interest is based mostly on the advantages of RFID as a technology for automatic identification in comparison with the classic barcode. However, its characteristics as a radio technology together with a potential combination of sensor technologies and low-cost information processing units make possible a range of applications for which the barcode appears to be completely unsuitable. Examples of the latter class of applications are the implementation of real-time location systems, the realization of new payment concepts (pay-per-use, pay-per-risk, etc.), and the combination of conventional products with online services. RFID systems can also provide an efficient means to prevent or delimit product counterfeiting. The technology overcomes several drawbacks of established brand- and product-protection measures. It allows for automated checking processes for products arriving in bulk, while at the same time offering a high level of protection against cloning attacks due to cryptographic measures and stable, user-friendly interfaces.

While the potential benefits of seamlessly linking objects and information systems are considerable (c.f. Fleisch et al. 2005), there are still several challenges to be overcome before a widespread, "ubiquitous" adoption of RFID may become reality. Besides the non-technical challenges such as the establishment of policies governing information sharing and access rights, privacy protection, agreements on data standards and models defining the distribution of costs, several hurdles regarding hardware and software issues persist.

In this chapter we discuss what brand owners and licit manufacturers can expect from RFID, and what steps can – or should – already be undertaken to reap the benefits. We start with a short excursion on Ubiquitous Computing technology as an enabler of the emerging "Internet of Things" and explain the role of RFID as an integral part of this development. Furthermore, we outline concrete solution concepts of brand- and product-protection measures based on RFID, and conclude with recommendations for their implementation.

9.1   An introduction to the Internet of Things

Ubiquitous Computing, Pervasive Computing, Ambient Intelligence, Silent Commerce – several different terms refer to a fundamentally new trend in electronic data processing.[48] All notions have in common an underlying vision of seamlessly augmenting ordinary objects with at least some fundamental means of information processing and data exchange. Following the ongoing miniaturization of electronic devices, advances in communication technology and the diffusion of global standards that facilitate the interconnection of such devices, one may already speak of an emerging Internet of Things in which smart objects are absorbed in – and fundamentally extend – the network of servers and personal computers we are familiar with today. Equipped with sensors and actuators, smart objects can respond to environmental conditions, thus making new production, logistics and service concepts possible. Applications go far beyond the frequently mentioned – and somewhat naïve – concepts of smart fridges that know when their content will expire and automatically reorder food. In fact, smart objects have the potential to profoundly change not only how future industry processes look but also how companies are managed and how cooperation among them is organized. The role of Ubiquitous Computing (UbiComp) technologies is probably best understood after briefly recalling the previous advances in organizational data processing systems. The following three models show how smart objects fit into this development.

Development phases of organizational data processing (model 1)

For a company or a network of companies the term "area of integration of information systems" refers to the number of tasks or processes that are supported by such systems. From the early days of electronic data processing the area of integration has steadily increased alongside the development of computer technology. The evolution can be broken down into 4 phases. In phase 1 companies aimed to support individual business functions (for example billing, demand planning, etc.) and set up insular computer systems. In phase 2 insular solutions within the most important departments were interconnected which, with the emerging software applications, allowed for an integration of interrelated functions (for example in accounting or purchasing). The potential (and limitations) of information systems influenced the design of the company's processes. Phase 3 can be characterized by the availability of advanced Enterprise Resource Planning (ERP) systems. They paved the way towards the integration of different departments and fostered efficient cooperation within the entire company. In phase 4 some companies started to establish systems that supported cross-company Electronic Data Interchange (EDI) with major suppliers and important customers, making mass transactions

---

[48] The following subsection is taken in part from the book "Das Internet der Dinge" (Fleisch and Mattern 2005).

more efficient. Customer orientation became more important, and downstream business processes increasingly influenced the way products and services were developed and brought to the market. Information systems today have to reflect not only the requirements of one company but also the needs of the company's customers. They have therefore developed – and are still developing – into better representations of the real business world. While integration is nearly complete with respect to both processes and collaborating companies, individual product instances are not yet sufficiently captured by information systems. Many business problems result from this shortcoming. They include out-of-stock situations in retailing, unreliable inventory data in warehouses, and wrong shipments (see Info Box 9.1 for two short cases). Moreover, many challenges with respect to customer-relationship management and one-to-one marketing exist. Ubiquitous Computing provides the link between individual items and computer systems. It can therefore be seen as the next step in organizational data processing.

Info Box 9.1: Two challenges due to the lack of item-level data

Stock-outs at the retail store are a major issue in today's retail environment. An ECR Europe study (ECR 2003) came to the conclusion that the number of stock-outs at the upstream echelons in the supply chain is low compared to the number on the retail shelves. Gruen et al. (2002) estimate that stock-out levels average 8.3%, where-upon the majority of stock-outs is not caused in the logistical chain from the manu-facturer to the store, but rather in the retail store itself. The authors estimate that wrong forecasting (13%) and ordering decisions (34%) are responsible for approxi-mately 50% of all stock-outs in store. Another major reason (25%), however, for retail stock-outs lies in the shelf-replenishment process, i.e. products are in the store, but not on the shelf (according to Thiesse et al. 2007).

Recalls appear to be an integral part of product launches in the automotive indus-try. In 2006 alone the German automotive industry initiated 167 official, large-scale recalls (KBA 2007). With information on what instance of a product has been built in which car, such measures could take place on a much more selective basis.

Integration of real-world data (model 2)

With increasing areas of integration, companies aim to avoid media breaks when collecting and processing business data. An example of a media break is the need for multiple entries of one order in several applications of a company's informa-tion system. For example, an order arrives by email, the order information is printed out and typed in the ERP system, the ERP system compiles a parts list that again has to be manually processed in the purchasing department. Media gaps often resemble missing links within an otherwise digital chain. Discontinuities of the flow of business data are often the reason for slow and error-prone processes.

Many companies have eliminated media breaks within their information systems whenever possible; the need to print out something just to re-enter it somewhere else is mostly a thing of the past. However, discontinuities are still very common when capturing data of physical events, for example when registering the number and type of goods received, the number of items stored in a warehouse, parts already built into a complex product, numbers and specifications of goods shipped, etc. Such data is captured manually, and, because the capturing processes are expensive, only at a coarse-grained level. Consequently the information is often not sufficient to support ambitious quality management programs, and even such fundamental data like quantities of items in a warehouse are not available at a desired level of accuracy. UbiComp technologies have to potentially bridge the media break between physical processes and information systems. They allow for an automated, real-time machine-to-machine interaction between smart objects and traditional computer networks and thereby act as a mediator between the real and the virtual world. Smart objects can, for example, without human interaction register themselves in a company's ERP system or keep track of the process steps they were already subject to. When describing the advancement of organizational information systems as an ongoing process of avoiding media discontinuities, UbiComp can again be seen as the next step in this development.

Improvements in data quality (model 3)

Improvements in the quality and availability of data can be seen as another source of motivation for the ongoing integration of IT systems. The rail industry may serve as an example to illustrate this development. Combining information from checkout with inventory data made efficient reordering procedures possible; granting the supplier real-time access to this information improved the process even further. The approach works as long as the inventory data is sufficiently accurate. However, human errors during checkout, wrong shipments, theft, etc. lead to deviations in the information system. Therefore companies have to conduct periodic inventory counts, trying to strike the balance between expensive manual labor for such measures and expensive stock-out situations. UbiComp technology can significantly improve the quality of the data. Improvements result from the increased quantity and timeliness of data (the variable costs of communication are extremely low, and measurements can be conducted frequently or even continuously), better location information (even low-cost devices can be localized if they are in close proximity to one of the readers), more fine-grained data on individual objects (identifying individual items rather than only their pallet), and further content such as data on environmental conditions during shipment, age, or handling instructions. UbiComp can therefore replace statistical analyses based on historical data with accurate real-time information.

The relationship of Ubiquitous Computing and RFID technology

Ubiquitous Computing is a collective term for highly miniaturized, embedded, networked microprocessor devices. They may come with sensors, actuators, extended memory, graphical user interfaces, and sophisticated radio modules. RFID transponders are a subset of UbiComp devices with a minimal set of functions: a communication module, memory for an identifying number and a microprocessor that coordinates the communication with reader devices. Thus RFID can be seen as a basic technology and as the first step towards more advanced devices. However, even transponders that can only communicate their identifier to a close-by reader device already allow the most significant media breaks to be avoided (see Info Box 9.2 for actual data). Moreover, a unique identifier can be translated into a Uniform Resource Identifier (URI), i.e. into a link to digital content, and thereby serve as an entry-point to the wide range of online-services or data repositories. The latter allows arbitrary amounts of data to be associated to a low-cost transponder.

The following sections deal with basic, low-cost RFID devices that are commercially available today. However, the reader should have in mind that brand- and product-protection measures only constitute one tiny fraction of what can be done with RFID, and that RFID itself is only the first step towards more sophisticated, smarter objects.

Info Box 9.2: Applications of RFID in 2007

In 2007, approximately 1.7 billion passive RFID transponders were sold worldwide. The total market including tags, readers, software, and services was on the order of USD 5 billion. The size per vertical market is estimated as follows (Das and Harrop 2008):

Tag location	No. of tags in 2007 (millions)	Tag location	No. of tags in 2007 (millions)
Air baggage	45	People (excluding other sectors)	0,8
Animals	80	Pharma, healthcare	0,3
Archiving (documents/samples)	8	Postal	1,2
Books	60	Retail apparel	95
Car clickers	47	Retail CPG pallet/case	225
Consumer goods	7	Shelf edge labels	0,1
Conveyances, freight	25,3	Smart cards, payment key fobs	630
Drugs	18	Tickets, banknotes, secure documents	250
Manufacturing parts, tools	40	Tires	0,1
Military	25	Vehicles	5,8
Other healthcare	12	Other	120
Passport page	45	Total	412,3

Privacy concerns among consumers[49]

Concerns about the possible risks of using RFID have increasingly been voiced. The risks associated with the technology include both the direct impact of electromagnetic radiation on human health, as well as indirect economic consequences such as the elimination of jobs through increasing automation. However, the most frequently expressed fear relates to the misuse of data generated by RFID, resulting in an undesirable intrusion into the privacy of individuals. Here, the fears of the general public extend from the analysis and evaluation of individual consumer behavior to an omnipresent surveillance in the form of transponder labels.

The debate has become additionally heated through the actions and campaigns of pressure groups such as the American Association "Consumers Against Supermarket Privacy Invasion and Numbering (CASPIAN)". For example, the well-publicized "Big Brother Award" given to the Metro Group, along with a demonstration on the 28th of February 2004 in front of the Metro Future Store in Rheinberg, caused them to ultimately withdraw the RFID-based customer cards that were in circulation at the time (c.f. Albrecht and McIntyre 2005). Further examples in Europe and the U.S., such as CASPIAN's call for a boycott of Gillette products because of tests with RFID transponders in razor blade packages, show that these are not isolated incidents. That this protest movement can bring about such sustainable effects, whilst working with the simplest of methods, permits conclusions regarding the significance that data protection and privacy have achieved in the populace as a whole.

In the conflict with RFID opponents, the retailers, producers, and suppliers of technology have so far taken a rather defensive, reactive position in the often heavily emotional debate. That is to say, the supply-side has adopted a restrained information policy and concentrated its arguments primarily on the technical aspects of RFID, with a strong focus towards a one-way, top-down process of consumer education (Givens 2005). As previous controversial technology-related topics have shown, this strategy harbors the danger of massive rejection on the part of customers, thus leading to a failure of the newly introduced technology.

The threat to privacy has its origins in the ability to permanently save and link information about individuals. Before the introduction of computer technology personal information about economic processes had no tangible value beyond individual transactions, and was therefore no longer used after them. This information has turned into a competitive asset – and it has become possible to formulate detailed individual profiles of customers and their purchasing behavior (Culnan and Bies 2003 and Spinello 1998). With RFID, yet another dimension to data acquisition has developed through (1) the temporal and spatial extension of data collection activities, (2) the inability to recognize and reconstruct data collection, (3) the acquisition of new data types through real-time monitoring, (4) the ever

---

[49] This subsection was printed by courtesy of our colleague Frédéric Thiesse. A more extensive analysis can be found in (Thiesse 2007).

decreasing transparency of reasons for acquiring data, and (5) the uncontrolled data access caused by extreme interconnectedness (Cas 2005 and Langheinrich 2005). In the case of RFID, the privacy-related problem arises particularly because of the globally unique identity of each good and the possible linkage with the owner. That facilitates, in principle, an automatic tracking of individual people (Juels 2005a and Sarma et al. 2003).

Mark Roberti, chief editor of the RFID Journal, once noted that "it doesn't matter how you plan to use RFID tags; what matters is how people think you may use them" (Roberti 2003). Starting from this perspective, the critical points from the public debate on RFID can be divided thematically into the following four categories:

- *Insecure technology.* The capabilities of the technology are in large part unclear to the people. However, it seems apparent that RFID implements inadequate functions for the guarantee of data security.
- *Unclear benefits.* The sense and purpose behind the introduction of RFID is not evident. This applies to the non-comprehensible benefits for companies, but above all the consumer has no obvious benefit from the technology for himself/herself. The misuse of customer data therefore appears to be the manifest area of application for RFID.
- *Deficient credibility.* The declarations made by commercial enterprises and producers are not believed. From the point of view of the consumer, these reserved information policies show that the reproaches of the consumer protectors and privacy activists are well-founded.
- *Inadequate legal position.* From the standpoint of many people the existing laws are not sufficient to provide protection against RFID for the individual. For this reason it is expected from lawmakers that they seriously limit the usage of RFID.

In summary, the observed discussions suggest that perceptions of RFID as a risk to the private sphere of individuals are dominated by massive fears and a deep-rooted mistrust of the firms that use the technology. It also becomes clear that a lack of information on the part of consumers is one reason, but by no means the only one, why this technology is rejected. On the one hand, insufficient action on the part of enterprises occurs at several levels, whilst on the other, pressure groups have succeeded in ensuring that some consumers have become conscious of the risks associated with RFID. For the most part, the public focuses on these risks and the potential benefits are suppressed or not perceived.

Enterprises are confronted with the question of what means they have at their disposal to influence risk perceptions in their favor. On the one hand, the objective must be to disseminate knowledge about the technology. On the other hand, where this is not possible and insecurity prevails, the development of a trust relationship which helps to surmount these insecurities must be facilitated. For this purpose we propose, on the basis of the previously described four core statements of the RFID critics, the following action levels as part of a strategy framework (see Figure 9.1):

- *Technology.* At a technical level, consideration should be given to expanding the functions of RFID components to make data misapplication impossible or at the least extremely difficult.
- *Processes.* At the process level the aim is to so configure the processes influenced by RFID that risks for the private sphere by default are reduced to a minimum and a more robust importance is given to elevating the benefits for the customer.
- *Dialogue.* The risk dialogue in and with the general public, as well as the individual consumer, is targeted at regaining lost credibility.
- *Rules.* Rules serve as the obligatory definitions for all sides regarding which applications and/or conduct in connection with the technology should be counted as permissible or not.

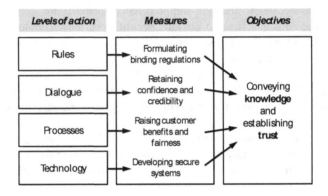

Figure 9.1: Strategy framework

The potential to secure data at the technical level is manifold and encompasses, apart from general IT security measures, a number of RFID-specific "Privacy-Enhancing Technologies (PETs)" as well. These prevent the uncontrolled reading of transponders as well as the manipulation of information saved in them. Some of these procedures can immediately be regarded as not viable, because they are simply impractical. This is a result of excessively high technical requirements or significant complexity for the user. On top of that, most of the solutions pass on to the consumer the organizational effort for privacy protection, similar to an opt-out procedure through the technology (Karjoth and Moskowitz 2005). From our perspective, the most serious problem seems that the additional security acquired is neither visible nor perceptible and, even worse, reliable verification is impossible. Therefore despite the clear need for additional and ongoing technological development, the objective of achieving an improved acceptance of technology cannot be achieved through such measures alone.

Another way of altering consumer attitudes towards RFID technology is the creation of incentives through changed procedures at the process level. On the one hand, processes should be formulated in such a way that customers gain the impression of "procedural fairness", that is, an appropriate handling of business activities (Culnan and Armstrong 1999). In this respect, apart from knowledge about procedures, being able to control them through, for example, opt-in choices, constitutes an important factor (Culnan and Bies 2003). In the context of a business relationship, opt-in refers to the requirement for someone to decide consciously in favor of a service, for example, by responding to a personalized advertising mail. Conversely, the frequently found and much less customer-friendly opt-out approach means that a positive decision is assumed in advance through presetting and the customer is obliged to reject this explicitly (Winer 2001). On the other hand, improved processes can increase customer acceptance of technology through additional services and benefits (Eckfeldt 2005).

An open dialogue with customers, independent of individual transactions, plays an important role in gaining or regaining confidence and credibility. In the specific context of RFID, strategies dominate that tend to understate risks or are oriented towards indoctrinating the public or just refusing communication. These are, however, communication strategies which are not particularly well suited to winning over customers for firms. Many of the following typical risk communication errors manifest themselves in the current debate: (1) denial and defensive information policy, (2) appeasement, (3) aggressive and confrontational interactions as well as polemics, (4) providing information too late, (5) reactive information policy, and (6) lack of clarity and comprehensibility of information. As a result of hardening fronts and consensus problems, it is difficult to achieve and develop a constructive dialogue. Nonetheless, the willingness to give interviews and practical demonstrations, cooperate with interest groups, use experts appropriately and so on are, over the longer term, successful measures and part of an open and proactive communication culture (Wiedemann and Hennen 1989).

A further instrument for the development of trust is the formulation of clear rules which are obligatory for all sides. For this purpose, two basic strategic options are available to enterprises, (1) an involvement in the official statutory process or (2) the agreement of industrial voluntary commitments. Whilst the integration of neutral statutory institutions brings a certain level of trust-bonus with it, voluntary commitments by industry have the advantages of faster reaction to the requirements of consumers, as well as informal control mechanisms. In both cases, it is important to know the existing data protection regulations when developing a strategy, especially considering the fact that fundamentally different approaches to the protection of the private sphere have established themselves in the USA and Europe over the course of the last few decades (Langheinrich 2005): The European approach favors comprehensive, all-encompassing data protection legislation that governs both the private and the public sector, while the sectoral approach popular in the U.S. favors voluntary industry regulations whenever possible, employing legal constraints only when absolutely necessary (Smith 2001).

Apart from government regulation, another option for the implementation of privacy-related rules is self-regulation (Swire 1997). For self-regulation to effectively address privacy concerns, organizations need to voluntarily adopt and implement a set of privacy policies as well as compliance procedures and enforcement mechanisms, so that consumers will have the confidence that an organization is playing by the rules (Culnan and Bies 2003). In some cases, trust in self-regulation may be fostered if there is an assurance to the consumer by a trusted third party that an organization's practices conform to the policies it has disclosed. Trade associations can also play a role in developing market solutions for privacy if membership in the association is conditioned on observing fair information practices. An example for this is the EPCglobal's Guidelines on EPC for Consumer Products (EPCglobal 2005). These commit EPCglobal member organizations to clearly show to the consumer the presence of an EPC-Tag in the product, to inform him or her of the possibilities for removing the tag, to make available further information on the functioning and application of EPC, and to guarantee that the EPC does not contain, collect or store any personally identifiable information.

With respect to an application of RFID to protect the consumer from counterfeit products, the objectives and benefits of the solution can be easily communicated. One can expect high acceptance rates among a wide consumer base, especially for goods that are often purchased as deceptive, potentially dangerous counterfeits.

9.2   Technical principles of RFID technology

So far we have simply introduced RFID transponders as a subgroup of UbiComp devices with a basic set of functions. Although it is not necessary for now to go into detail down to bit level, a good understanding of the basic operating principles is helpful to anticipate the technology's limitations as well as potential hurdles during adoption. Therefore we will briefly discuss how RFID transponders work and what infrastructure facilitates the data exchange.

RFID transponders and readers

Although RFID systems which exploit surface acoustic wave delay lines or resonant circuits to transmit an identifying signal exist, the vast majority of RFID tags – and the class of transponders which is relevant in the context of this book – utilize silicon-based microchips to store data (for example a serial number) and to establish a data exchange with a reader device. The energy for their operation stems from an internal battery (for active or semi-passive tags) or is retrieved from the electromagnetic field generated by a reader device (for passive tags). In general battery-powered tags have wider read ranges, meaning that the distance between reader and transponder can be greater than with passive devices, but they are more expensive, larger in size, and have a shorter lifetime. Active or semi-passive tags are often used in closed-loop systems, for example for container tracking, or for data-logger applications, as their battery can continuously supply sensors and the digital logic. Most supply chain applications as well as the anti-counterfeiting approaches that will be presented in the next section are based on passive transponders. They are cheaper to manufacture, smaller in size, have an almost unrestricted lifetime as they do not depend on batteries, and are more straight forward to deal with during recycling.

However, using the reader device for energy supply can be a bit tricky. The minimum field strength under optimal conditions that is required to power the tag depends on its minimum operating voltage, the frequency of operation and the effective area of the antenna. Minimum operating voltage depends on the design of the tag (it is in fact an important measure of the silicon chip's quality); frequency of operation is usually chosen in accordance with established standards (and the choice of standards is in turn influenced by the object the transponder has to be attached to); and the effective area of the antenna depends on its actual size and the design.

In practical settings where (1) antennas may not be in unobstructed space, (2) impedances may be mismatched (that is the antenna is not exactly adjusted to the frequency), and (3) antennas may not be correctly aligned and polarized, the maximum read range can be significantly decreased. The primary effects are absorption and reflection by surrounding dielectrics and conductors, multipath fading which can lead to variations in the received signal strength in a relatively short period of time, polarization losses, and the influence of the material surrounding the tag on the antenna's impedance. As a consequence the maximum read range is reduced and transponders have to be closer to the reader than under laboratory conditions. Besides the physical characteristics of the radio channel, maximum read ranges and data-transfer rates are further restricted by radio regulations which confine bandwidth and transmission power to limit interference with other radio devices.

So what is the problem with transponders that are not sufficiently powered? Well, they do not show up on automated inventory counts. Missed reads are at odds with the improved data quality we were aiming for. For brand- and product-protection measures they are even worse, as one may have to ask if a product has not been identified due to technical problems or because it is not a genuine article.

However, with carefully designed RFID systems, the above-mentioned challenges can be met. We spent much more time on the challenges that need to be overcome when developing an RFID-based solution than on the shortcomings of other security technologies. Does this mean that RFID systems are less suited for

---

Info Box 9.3: Frequencies and read ranges

RFID systems can be classified based on the frequency band that is used for the communication between tag and reader. Typical frequencies are: 125–135 KHz (Low Frequency, LF), 13.56 MHz (High Frequency, HF), 868-930 MHz (Ultra High Frequency, UHF), and 2.45 GHz (Microwave, MW). The different frequencies have characteristic advantages and disadvantages with respect to the maximum read ranges, performance in proximity of dielectric and conductive material (e.g. water and metal), size of the tag, and cost of tags and readers. Passive transponders either realize coupling through the near field (LF tags and HF tags), or far field (UHF tags and MW tags). Near filed systems base on inductive coupling between readers and tag antennas through a changing magnetic field. Thus, the transponders' antennas resemble coils in a transformer system. With LF and HF systems, maximum read ranges of typically less than one meter can be realized. The performance of LF and HF systems in general is less susceptible to the presence of dielectric and conducting objects in the tag vicinity. At LF and HF, the electromagnetic regulations are enforced in the far field and only allow minimal radiation. However, since the near field energy density per unit volume decreases as the inverse sixth power with the distance from the antenna, substantial energy densities can be obtained close to the reader. This enables energy-intense, complex cryptographic operations (e.g. to authenticate a transponder) in close proximity to the reader. Far field (i.e. UHF and MW) systems use electromagnetic waves propagating between reader and tag antennas. A dipole antenna in the tag is used to retrieve the energy from the field. Under good conditions, far field readers can successfully interrogate tags that are up to 4 to 10 meters away. Corresponding antennas are often less expensive than the coils of LF and HF systems. However, the antennas often have to be customized to the products the tags are attached to, and computationally intense operations on the tags are difficult to implement due to energy constraints.

---

supply chain security applications? Quite the contrary. When trying to satisfy the requirements we mentioned in Section 8.3, there is little point in dwelling on the deficiencies of the established security approaches. Optical and chemical features lack the standardization that would facilitate user-friendly inspections across different product categories and do not support automated, large-scale inspections of goods arriving in bulk. With RFID systems this becomes possible. When outlining the challenges, we just want to make the point that solution design is not always easy. While off-the-shelf products might work in RFID-friendly environments, for example for most plastic material, glass or textiles, extensive testing is still required for objects that are made of or contain metal or water, for extreme line speeds, and for items where a customized transponder is already embedded in the production process.

Info Box 9.4: NFC enabled cellular phones[50]

Near Field Communication (NFC) denotes a short-range wireless connectivity technology that evolved from a combination of existing contactless identification standards. NFC enables cellular phones to act as RFID readers and emulate RFID transponders. The devices operate at 13.56 MHz (i.e. in the HF band) and have typical read ranges of 10 to 20 centimeters. While EPC technology is primarily seen as a solution to support supply chain processes, NFC is specifically tailored to consumer applications. The devices are compatible to the following international standards:

- ISO/IEC 18092 (also referred to as NFCIP-1),
- ISO/IEC 14443 (smart card technology, "proximity coupling devices"), and
- ISO/IEC 15693 ("vicinity coupling devices").

NFC allows for an interaction with a large number of already deployed application infrastructures. Prominent domains include mobile payment, mobile ticketing, the transfer of data from one device to another, and the download of information either from other NFC devices or, using the connectivity of cellular phones together with address information stored on the transponder, from online sources (c.f. ecma international 2005, innovision 2006, NFC Forum 2006, and GSM Association 2007). Many NFC applications have already been deployed in pilot projects. Examples include ticketing in Xiamen, China and Hanau near Frankfurt, Germany. Two further pilots are planned with the New York public transportation system and the London Tube (Hargrave 2006). Contactless payment services constitute another important application domain. Solutions are being introduced by Visa ("VisaWave"), MasterCard ("PayPass"), and American Express ("express pay"). By the end of 2006, credit card companies had shipped 20 million contactless cards (compatible to ISO/IEC 14443) which can be used with 205,000 readers at 45,000 locations (Mullagh 2006). NFC enabled phones have been introduced by Nokia. They are well suited for anti-counterfeiting applications. A drawback, however, is that the NFC devices are not compatible with EPC transponders.

Infrastructure and standardization

In most practical settings the value of RFID applications stems from an interconnection of multiple readers, databases and information processing units. Compiling track-and-trace data while an item is moving through the supply chain, for example, requires different parties to record arrival and shipping dates in a way that allows for collecting the data whenever necessary. This in turn calls for an agreement upon some sort of standards for data representation and data exchange

---

[50] This Info Box was printed by courtesy of our colleague Thomas Wiechert. A more extensive analysis can be found in Wiechert et al. (2007).

among the stakeholders. Fortunately such standards are not only required for the design of anti-counterfeiting solutions but constitute a prerequisite for almost every cross-company application of RFID systems that we outlined in Section 9.1. Addressing this issue, the predominant RFID standardization body, EPCglobal, has published a set of data interfaces and interrelated standards for hardware and software with the aim of increasing visibility throughout the supply chain and facilitating communication between different entities. The corresponding network infrastructure is termed the EPCglobal Architecture Framework,[51] also referred to as the EPC Network, which in turn is named after the standardized electronic product code (EPC). Besides tag, reader and data exchange protocols, the pivotal components of the framework are Lookup and Discovery Services and the EPC Information Service (EPCIS). Their basic characteristics are briefly introduced below.

The Electronic Product Code (EPC)

The Electronic Product Code (EPC) is a numbering scheme that is used to identify objects. It incorporates existing EAN.UCC keys and U.S. Department of Defense constructs. Its fits into 96 bit of memory and is thus well suited for low-cost transponders, while the numbering space is sufficient to identify individual items (i.e. all instances of a product) instead of manufacturer and product class only (see Figure 9.2). An EPC-compliant tag contains a unique EPC number. The EPC was developed by the MIT Auto-ID Center (now Auto-ID Labs) in the late 1990s. The EPC system is currently managed by EPCglobal, Inc., a subsidiary of GS1 which defined the UPC barcode.

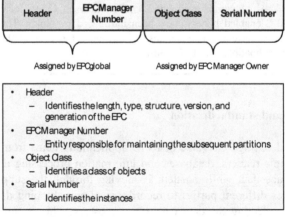

Figure 9.2: Basic format of an EPC code

---

[51] For readers with a background in information system design, the framework document is certainly worth reading (EPC ARC 2005).

Discovery Service

Let us assume that a pharmaceutical company allows its customers (or customs officials) to check online if a given blister pack has been produced by the company and if it is intended for sale in the actual geographic market. The item carries a simple, low-cost RFID transponder with a unique identification number in a standardized format. The customer is equipped with an RFID reader device with online access (for example an RFID enabled cellular phone). With this device he or she can retrieve serial numbers from RFID transponders that adhere to a given standard, regardless whether they are affixed to car tires, handbags or pharmaceutical products. But how does the device find the online address of the authentication service that is provided by the pharmaceutical company (and later, this of the tire manufacturer)? Writing the access information on the tag is not necessarily the best approach as it may change over time, and its owner cannot update the abundance of transponders that might be in circulation by then. To ensure that users (or systems) can conveniently retrieve the addresses of other (decentralized, potentially previously unknown) entities, the logically localized Lookup Service is implemented. In principle the address lookup system allows a manufacturer to register a data pair consisting of a serial number and the corresponding network address. The lookup service translates a unique identifier (an EPC) into the address of the network resource of the corresponding EPC Manager (i.e. the entity that assigned the EPC to the object; typically, the manufacturer) where further information (for example for an authenticity check of the corresponding item) may be retrieved. It can resemble the Internet Domain Name Service (DNS) and can be implemented as a hierarchy of lookup services. In such a setting the reader device only needs to know the address of one service rather than all potential addresses. Only registered manufacturers should be allowed to update the lookup service as this will prevent illicit actors from inserting bogus redirects, or, if this is not possible, the authenticity of the anti-counterfeiting service provider has to be ensured by other means. For the user the procedure is completely transparent, comparable to clicking on a link in a conventional Internet browser. The standard is described in EPC ONS (2005).

EPC Information Service (EPCIS)

The EPCIS can be seen as the core component of the EPC Network. It covers interfaces for information exchange as well as data specifications to ensure that, for example, all parties provide information on a product's history in the right format. The data can be divided into static and transactional information. Static information does not change over the life of a product. Examples are specification of the product class, date of manufacture, lot number, etc. Transactional data may be updated or extended over the life cycle; it can be divided into (1) instance observations including time, location, associated EPCs, and business process steps (for example "the product X was aggregated to a pallet at building 3 at 10:43pm"), (2) quantity observation including time, location, and object class information (for example "there are 12 razors on the second shop floor at 12:23pm"), and (3) business transaction observations including time, associated EPCs, business process steps, and business transaction identifiers (EPC ARC 2005). Transactional data which relates to one article is often generated by several enterprises – including for example its manufacturer, a logistics service provider, a wholesaler, and a retailer – and stored in their own EPCIS systems.

In order to gain a higher level of supply chain visibility, the business partners have to assemble the distributed information. The first step is to determine where the relevant information can be retrieved. This may be an easy problem when the addresses of the supply chain partners are known in advance, for example in two-party scenarios or in a static, predefined business relationship. However, it is a major challenge when more, potentially unknown enterprises are involved. One approach, referred to as "following the chain", is based on each business partner's knowledge of the address of the succeeding supply chain partner and relies on the cooperation within the chain to pass the query through to the manufacturer. A drawback is the potential lack of cooperation. A second approach is based upon an application termed EPCIS Discovery Services which locates the EPCIS services that have information about the object in question. Compiling the numerous data points into meaningful information while concealing confidential information and protecting the data record from unauthorized access constitutes a major challenge for the design of the EPC architecture. We will come back to this issue when discussing track-and-trace-based anti-counterfeiting solutions in the following section.

Info Box 9.5: Further reading

Infrastructure: More information on the EPC Network and its standards is available in the EPCglobal RFID cookbook. It contains a cost tutorial, practice briefings, information on pilot studies, and information on compliance certification (www.epcglobalinc.org/what/cookbook/)

Transponders: Our companion book "Networked RFID Systems and Lightweight Cryptography – Raising Barriers to Product Counterfeiting" provides an in-depth discussion of the technical aspects of RFID-based anti-counterfeiting systems, with a focus on the transponders and hardware (Cole and Ranasinghe 2008). A thorough discussion on the technical principles of RFID systems can be found in *The RFID Handbook* by Finkenzeller (1999 and 2006).

Applications and consumer information: Different applications are outlined under *discover rfid*, an explanatory website run by EPCglobal. The site also features links to research projects as well as some information on privacy-related issues (www.discoverrfid.org/).

9.3   Solution concepts

In this section we introduce three RFID-based anti-counterfeiting solutions that vary in cost, complexity and level of security. All approaches allow for automated inspections of products arriving in bulk and offer stable and easy-to-use interfaces that do not depend on the underlying security solution. However, each approach comes with some challenges that need to be considered before an implementation.

Plausibility checks based on unique serial numbers

Marking individual items with unique identifiers helps to keep track of goods and thereby to detect illicit trade activities. In its basic form only the production event of each item is captured. The operating principle of a respective system is quite simple. The manufacturer generates a (random) [52] identification number, saves it on a data carrier that is attached to the item, and stores the number in a database (possibly together with the production date, the specification of the associated item, etc.). During the verification process, a reader device retrieves the ID from the data carrier and sends it to a service offered by the manufacturer which looks up the number in the database; if the number is found to be valid (i.e. if it has been registered and the item is indeed available for sale), this can be interpreted as proof of the authenticity of the product. In an ideal scenario,

- the validity of the number can be easily checked by the supply chain partner or, if desired, by the consumer,
- the number can be read automatically by authorized individuals, allowing for large-scale searches for bogus identifiers,
- it is hard to simply guess valid IDs,
- a duplication of the number carrier is unreasonably expensive,
- the number carriers cannot be removed nor can illicit actors overwrite them as this would allow the identity of the object to be disguised, and
- the system does not enable unauthorized parties to draw conclusions about production volumes, target markets, distribution strategies, etc.

With a carefully designed system one can come very close to the desired characteristics. Therefore the following aspects should be considered.

Selection of an address service. Product checks may have to be conducted at various places, for example in stores or at street sales. If only a known group of investigators conducts such inspections or if the number of user devices is limited, the

---

[52] The number, however, should not be truly random as it has to be made sure that all assigned identifiers are assigned only once. A solution could be to concatenate a unique manufacturer and product ID together and a number for each instance that is chosen (without repetition) out of a large, reserved numbering space.

network address to access the database can be made available in advance, for example by providing it to the service personnel or by directly programming it into the test devices. However, if the devices are used by changing actors, if the test equipment is not under the control of the brand owner, or if many brand owners use the same approach, a "hard-wired" solution is not feasible, and a standardized, trusted address lookup system is required. A suitable design (the Lookup Service) was described before, where the reader retrieves the serial number (the EPC) from the tag, sends it to the address lookup system, and gets back the address where further services are available. Furthermore, as the EPCs are managed by one organization, it is ensured that numbers are not assigned more than once. We highly recommend taking advantage of this standardized approach as it allows inspectors and consumers to use one device, maybe even an RFID enabled cellular phone, for various products. This is what makes product checks convenient (and thus feasible), and, in fact, what makes RFID-based brand- and product-protection systems superior to other approaches.

Protection against number guessing. Illicit actors can try to guess an identification number, hoping that the identifier is valid and has been assigned by the manufacturer to the targeted product. However, if IDs come from a numbering space that is significantly larger than the number of items, it is extremely unlikely that an illicit actor is successful in finding a valid ID[53] – as it is, for example, unlikely to just guess a long password. However, the brand owner has to make sure that the numbering space is used properly. Counting from one or using a predictable algorithm to choose IDs would not be a good idea. Moreover, it should not be made easy for illicit actors to test many IDs to find a valid one. The latter issue is addressed in the next paragraph.

Protection against unwanted access/espionage. Problems occur if unauthorized parties test a large amount of random numbers and find out which IDs are validly registered in the database. They could use information not only to fraudulently mark counterfeit products but also to draw conclusions regarding production output, market share, etc. Therefore a mechanism should be implemented to restrict number guessing by third parties. Suitable approaches are user access rights management, throttling, or partial disclosure. User management is well suited when the inspectors are known (for example when inspection processes are conducted only by designated employees). In non-predetermined supply chains with a large number of trading partners, access right management is likely to be impractical if the registration requires some effort on the part of the user. Since the incentives for conducting product checks are often limited, additional user interaction may significantly limit the acceptance of the system. We suggest using a weak form of authentication, for example by registering caller IDs or the IP addresses of the client. With this information the system can detect excessive queries and restrict

---

[53] That is many numbers are not assigned and thus invalid.

access accordingly. The intentional restriction of the number of answered queries during a defined period of time is referred to as bandwidth throttling. Another approach is called partial disclosure; it can be described as the fragmentary comparison of ID numbers with the contents of a database. The lookup system may require the reader to transmit a complete ID number, but only compare a defined set of randomly chosen bits with the database system. Even if an attacker receives an acknowledgement for a chosen ID, he or she cannot be sure that the number is correct as the result only refers to a (unknown) group of bits. Since the next query would very likely check for a different subgroup, the result does not notably help to compromise the system.

Protection against duplicated transponders. Basic RFID transponders do not require authentication on the part of the reader; they allow every compatible device to retrieve the ID that they carry. Since transponders with field-programmable memory are available, this enables illicit actors to read out the memory from licit transponders, program them to other tags, and attach these duplicates to bogus articles. This is in fact a nightmare for any security technology provider. In order to provide some protection against cloning attacks, it is desirable to implement a feature which cannot easily be copied. Fortunately transponders which have been produced to date contain a unique, read-only Tag ID which is set during the manufacturing process (similar to PC network cards, which also have a unique hardware address). Using the number pair (Serial Number/Tag ID) instead of the serial number only results in a considerably more secure solution without additional hardware expenses. Consequently RFID-based anti-counterfeiting systems should also include the Tag ID in the verification process. Moreover, under certain conditions it is possible to detect duplicate serial numbers, for example when they are read out at disparate locations within an unreasonably short time, when one ID is checked unreasonably often, or when they are read in untypical geographic markets. Credit card companies lived well with similar business logic for quite a while – and credit cards resemble very high value goods in terms of the potential losses that can result from duplicates. Now, following increasing numbers of fraud cases, these companies are moving on to more advanced, electronic features which are comparable to what we will introduce later. Brand owners could do the same as soon as there is a positive business case for the more expensive solutions. With RFID they can do this while keeping a considerable part of the existing infrastructure and without changing the way users interact with the system.

The fact that low-cost transponders can be cloned is a major shortcoming of this approach. For product categories where substandard imitations are likely to have severe implications for the consumers' health and safety, such features may not be sufficient alone; we describe some more secure approaches below. However, for goods where counterfeiting results in incremental financial losses, this basic approach is well suited as it facilitates large-scale inspections throughout the supply chain.

Plausibility checks based on track-and-trace

In track-and-trace systems information on an object's location and the corresponding time of the identification, possibly together with data on its owner and status, is recorded and stored for further processing. If such measurements are repeated over time, they allow for plausibility checks of the product's history. Heuristics can be applied for example as done by credit card companies which routinely block cards if they have a suspicious transaction history.

Track-and-trace systems rely on the ability to uniquely identify individual articles. In order to conduct meaningful analyses, numerous data points have to be collected, which requires an efficient way of capturing supply chain events. In this context the unique serial number approach as outlined before is an enabling technology, or, in other words, can be seen as the first step towards a track-and-trace solution.

Though the operating principles of track-and-trace systems may appear simple, an actual implementation of the infrastructure is a severe challenge. From a technical perspective access management in non-predetermined supply chains in particular constitutes a major hurdle. Detailed information on the flow of goods, quantities shipped to individual wholesalers, number of articles sold into different countries, choice of logistics service providers, etc. are extremely well-protected secrets. Track-and-trace data could reveal sensitive details and thus has to be protected, and doing this may not always be easy. However, even bigger barriers seem to be organizational issues on the ownership of the data, the distribution of system costs, and the lack of interest among some parties to provide their customer with a high degree of supply chain visibility. Without a powerful regulatory body that mandates the use of such systems, it is unlikely that enough companies will participate in such a collaborative approach, and a consistent product history for every item will not be compiled.

Nevertheless, track-and-trace is an important and promising approach, for example for the pharmaceutical industry. The obstacles that we have pointed out are to be seen alongside other related benefits such as enhanced inventory management, production and distribution control. However, for the majority of brand owners and without governments or strong industry associations pushing for a solution, such systems are not likely to become a reality in the next three to five years. We rather expect history-based security systems to slowly evolve alongside the adoption of EPC technology with its focus on supply chain efficiency.

Object-specific security

Security solutions which are based on tagging technologies have a system-specific drawback: When checking an object, it is still the tag (for example a hologram or the RFID transponder) which is authenticated and not the object the tag is attached to. In other words, the link between tag and object is often not provided. In theory, and also in practice if the solution is not designed properly, a tag can be removed from an original article and attached to another object, thereby compromising the security system.

In contrast to most other tagging technologies, RFID can overcome this shortcoming. Even low-cost RFID tags with a certain amount of memory can store data that binds a tag to a given product, as a picture in a passport binds the document to its holder. An exemplary data set, termed product validation data, is given below.

Product Validation Data := {
      Unique Tag ID,
      Unique Serial Number,
      Product Specific Data,
      Signature Method,
      Signature Value };

This information resembles the picture or fingerprint in the passport analogy. The data has to be characteristic for an individual object, stable over time, and easily measurable during inspection. Which properties may be selected depends on the specific physical, chemical, electrical, etc. properties of a given object. Examples of characteristics are, either altogether or as a subset thereof, weight, electric resistance, form factors, a serial number printed on the product itself or its packaging, etc. This data will typically be written on the tag by the product's vendor before product delivery, for example during packaging. It is also possible to store a reference to the data on the tag, such as a URI that specifies an entry in a remote database. This may help to save tag resources, but will make product validation dependent on the availability of network connectivity.

An advantage of this approach is that low-cost tags with approximately 32 to 64 bytes of memory can be used. It does not rely on cryptographic functions on the tag, which would require more expensive transponders. The approach can also be combined with plausibility checks based on track-and-trace or secure tag authentication principles to avert cloning attacks.

Cryptographic authentication

As outlined before, basic RFID transponders do not require authentication on the part of the reader and consequently allow every compatible device to retrieve the ID that they carry. An advanced attacker may be able to obtain an identifier from a tag and program it into another transponder, or emulate the tag using other wireless devices. If done on a larger scale, duplicate devices can render track-and-trace solutions ineffective.

Challenge-response protocols can avert tag cloning as they allow for a comparison of secret keys at disparate locations without transferring them over a possibly insecure channel. The concept is by no means new. Corresponding mechanisms are well established in network security and in RFID-based payment solutions. In a carefully designed system third parties are prevented from reconstructing the secret, even if it is used numerous times. These properties qualify the technique for application in an RFID environment, where the channel must be regarded as insecure.[54]

Info Box 9.6: The principles of challenge-response mechanisms

A challenge-response protocol is a security mechanism that allows a person or an electronic device to verify the identity of another entity. The entity that is to be authenticated must prove that it knows a secret without sending it over a potentially insecure channel as this could allow eavesdroppers to capture the secret and mimic the authentic entity. The protocol fundamentally depends on the existence of one-way hash functions. These mathematical functions take an input and return the so-called hash value that only depends on the input. With a good hash function, the reverse calculation is "computationally infeasible", which means that determining the input from the output is so difficult that even high performance computers cannot find the solution in a reasonable amount of time (e.g. within the lifetime of a product series). As an example, let us assume that an RFID transponder should be authenticated by a server. The server and the genuine transponder know a shared secret (the key) that has been generated when the transponder was initialized and that has been stored in the manufacturer's database as well as in a read-protected memory block on the transponder. No other party knows the secret. To verify that the transponder is the device it claims to be, the server sends a challenge (typically a random number of a defined length) to the transponder. The transponder takes the random number together with the secret key as an input to calculate the hash value. This value is then sent back to the server, which also derives the result of the hash function based on the challenge and the secret key. If the result of the transponder matches the result of the server, the server can assume that the transponder knows the secret key and therefore is authentic. During this process, the key is not transferred over the channel and since the server can vary the challenge, replay attacks are not possible.

Challenge-response protocols require the transponder to perform quite complex calculations, and critics of this approach frequently mention the increasing tag costs which may result from the integration of the computational logic. However, the increase is only marginal given that the production volume would be the same (c.f. Info Box 9.7). Challenges are more likely to result from energy consumption and bandwidth constraints, as we will see below.

---

[54] A number of interesting low-cost cryptography issues are discussed in Juels (2004, 2005b, and 2006). See also Feldhofer et al. (2004) and Feldhofer et al. (2005) for an implementation of the Advanced Encryption Standard (AES) in resource-constraint RFID devices.

Info Box 9.7: Cost factors during tag production

When this book was printed, a 96-bit UHF EPC inlay, i.e. a chip and antenna mounted on a substrate, cost from 7 to 15 U.S. cents. LF and HF tags are slightly more expensive. The price covers the silicon microchip, the chip's packing, the external antenna, assembly, and the margin of the producers. In fact, the silicon chip is only one of the cost drivers. Therefore doubling the chip area to incorporate more memory or to build in an advanced arithmetic unit does not necessarily double the cost of the transponder. The reason that more advanced transponders are significantly more expensive results mostly from smaller production volumes.

In the following the mode of operation of a challenge-response protocol is described. An attempt is made to outline a minimal solution in order to facilitate further evaluations rather than to develop a real-world protocol. Only the commands "Write to Tag" and "Read from Tag" are required for communication between tag and reader during authentication. Consequently the proposed rudimentary protocol does not require any changes in existing reader protocols, the middleware, or the communication infrastructure for most established RFID systems. It nevertheless allows for an assessment of the implications on important system parameters beyond hardware costs and helps to reveal several obstacles of challenge-response mechanisms in low-cost RFID systems.

The principle communication architecture is outlined in Figure 9.3, and the tag is illustrated in Figure 9.4. Components of the device are – besides the RF interface, memory for Tag ID and serial number – three more registers, denoted status and challenge (both read-only for external devices), response (write-only), a crypto unit, as well as a register to store the secret key. All registers are 128 bits wide, resulting in an EEPROM size of 64 bytes excluding Tag ID.

Tag initialization. The secret key[55] is generated by the manufacturer, stored in its internal product database and transferred to the transponder in a protected environment. After the key is loaded to the corresponding register, the signal path connecting it with the data interface is destroyed and the key can no longer be read from external devices. From now on, the key is only accessible by the authentication service and the tag's cryptographic unit. It will not be transferred over an insecure channel during the authentication process.

---

[55] We use a symmetric key approach as they are less computationally intense and thus better suited for low-cost devices.

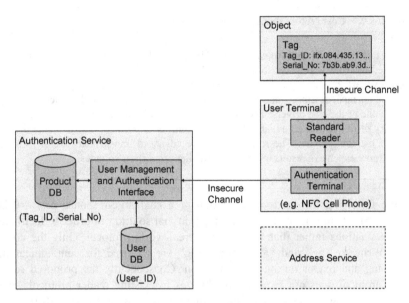

Figure 9.3: The main components of the authentication system

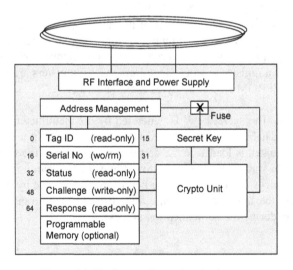

Figure 9.4: The layout of an authentication tag

The authentication process is initiated by a user terminal after reading out the tag identifier and looking up the network address of the corresponding authentication service.[56] The challenge is generated by the authentication service hosted by or on behalf of the manufacturer and sent to a specific register of the given tag upon request. When this happens the cryptographic unit is reset, the status register is updated so that it represents the ongoing authentication process, and the content of the challenge register together with the secret key are used to calculate the response. The result is then written into the response register where it can be read out from the reader device, and the status register is updated again. After completion of the calculation the response is sent to the authentication service, which checks its validity and returns the result of the query to the user terminal.

The crypto unit can make use of established and reviewed cryptographic primitives such as AES. An implementation is also possible with commercially available tags, such as the Mifare series, although these tags may use other mostly proprietary encryption algorithms. The protocol is outlined below. Most of the commands are chosen to be self-explanatory. The RFID reader device uses the following commands to communicate with the tag:

```
Result := Read_from_Tag (Select_Tag, Address),
Write_to_Tag (Select_Tag, Address, Data),
```

where Select_Tag specifies the tag for the operation and Address selects the start of the memory partition at which the Result is read or Data is written.

Tag authentication protocol:

```
<Read Tag_ID and Serial_No>
<Retrieve Address of Authentication_Service for Serial_No>
<Initiate communication between Authentication_Service and Authentication_Terminal>
<Authenticate Tag:
    Authentication_Terminal_request_Challenge (Tag_ID, Serial_No);
    (Serial_No, Challenge / Tag_not_valid) := Authentication_Service_return_Challenge;
    Authentication_Terminal_to_Reader (Write_to_Tag (Serial_No, Challenge_Address,
                                                                   Challenge));
    Authentication_Terminal_wait_until_Status_Register_signals_completion;
    (Response) := Authentication_Terminal_to_Reader (Read_from_Tag (Serial_No,
                                                                   Response_Address));
    Authentication_Terminal_requests_Response_check (Serial_No, Response);
    ("valid" / "not valid") := Authentication_Service_return_Result>
<Close_Connection>
<Display Result>
```

---

[56] In an actual implementation that also allows for non-crypto tests, the authentication service has to communicate which method (unique ID only, object-specific security, or challenge-response) is supported by the actual transponder. However, as these steps do not affect tag cost and tag-to-reader communications, they are ignored here.

Besides the potential seamless integration into an existing infrastructure, one advantage of the proposed solution is that the secret key does not have to be distributed after it is written to the tag and stored in the database. The reader devices neither have to know the key nor have to support other capabilities other than read and write. However, the protocol also highlights several drawbacks of challenge-response authentication for passive RFID solutions. They are related to:

- *Communication bandwidth.* Rather than just identifying the transponder, i.e. reading its serial number, an authentication process comprises several steps: (1) identifying the transponder, (2) addressing the individual transponder and the defined memory bank in order to (3) write a challenge to its memory, (4) addressing the tag and (5) checking its status bits for the completed calculation, and, after potential repetitions of step five, (6) addressing the individual transponder and the defined memory bank in order to (7) read out the response. Neglecting the communication overhead and assuming word lengths of 96 bit of the tag serial number, 128 bit for the challenge and the response, 32 bit for addressing individual transponders, and a width of the status register of 8 bit, 456 bit of data are to be transferred, compared to just 96 bit (again without overhead) of the simple ID approach which is sufficient for example for track-and-trace systems. When the application scenario requires bulk reading, several tags have to share a common communication channel with a restricted maximum available bandwidth. Consequently the number of transponders which can be authenticated within a given time or at a given conveyor belt speed is considerably below the bulk reading performance of simple ID systems.
- *Power supply.* Generating the response is computationally intense. Even low-power implementations of crypto-tags require much more energy than simple ID tags. For sophisticated operations where transponders not only dissipate more power but may also have to stay activated for a longer period of time to complete a calculation, fluctuations in energy supply can dramatically reduce read rates and consequently the maximum read range.
- *Tag costs in real-world applications.* Given the attack model introduced earlier, possible hardware attacks include power analysis and destructive attempts to access otherwise read-protected registers from the cryptographic unit of the integrated circuit. Though extremely difficult, such attempts could help illicit actors to uncover the secret keys that may be used in the authentication process. In order to avert these attacks, attempts are made to disguise power dissipation and to obscure the data paths and register alignments by introducing additional power sinks and using less regular design structures. This leads to higher power consumptions and increased chip sizes with implications for read ranges and transponder cost.

Again, we spent much time on the potential hurdles of RFID-based anti-counterfeiting measures. And again, we did this not because we think solutions are poor but to help readers assess applicability in industrial applications. RFID transponders with advanced cryptographic features are not yet cheap enough for cost-efficient application for low-cost goods, and they offer only limited bulk-reading capabilities at shorter read ranges. Track-and-trace applications require extensive cooperation among various supply chain partners, and data sharing and data access rights may lead to caution among the potential participants. Solutions that solely rely on unique tag IDs potentially become subject to cloning attacks. However, the level of security that even low-cost transponders offer is sufficient for most product categories, and the technology facilitates convenient, automated, large-scale checks of products arriving in bulk. Track-and-trace solutions may be difficult to establish, but if such documentation systems are mandated, RFID can help to cost-effectively collect the required data. Security tags with challenge-response mechanisms are highly secure and thus suited for protecting critical parts, for example in the automotive or aviation industry. Limited-bulk reading capabilities may not even be an issue for the high-value goods, read ranges will be improved with more advanced transponder design, and cost will fall considerably when larger quantities are demanded.

All approaches are built upon the same infrastructure and all can utilize the same test equipment. Migration paths towards more advanced solutions are at hand. RFID-based product inspections do not require any knowledge on the underlying security technology and improvements do not automatically lead to changes in the test procedures or user interfaces. Large-scale checks are feasible, and involvement of consumers in product-security measures is possible. Relying on low inspection rates due to intricate and time-consuming features therefore becomes less promising, which can dramatically reduce trade in illicit products. We will describe different application scenarios and their implications for both counterfeit producers and brand owners in the following section.

9.4 Application scenarios

RFID has the potential to affect counterfeit trade in various ways. To assess the implications in greater detail, three application scenarios – an insular approach, a solution to support customs, and an application based on track-and-trace – will be investigated below. Thereafter we outline the implications for monitoring, reaction, and prevention activities and discuss the effects of RFID for the different types of counterfeit producers.

Application scenario I: An insular solution

The first use case describes an application of RFID within a single company. Transponders are integrated into individual, exclusive, and frequently counterfeited luxury goods. Product checks take place at retail stores and are conducted by a dedicated inspection team.

Production steps. For the product under study low frequency (LF) transponders are integrated at item-level during an injection molding step; though ultra high frequency (UHF) transponders that could be attached to individual packages provided higher read ranges at lower tag costs, the in-product solution is preferred since individual articles are frequently repacked on the shop floor. Moreover, the integration makes removal and re-use of the transponder for a potentially counterfeit product very difficult. Tags which can be written only once ("write-once-read-many" transponders) are used. During production a 128-bit random number is generated and stored on the tag. The unique Tag ID is stored in a database together with an additional shorter random number that is written on (or engraved in) the product, information on the product type, the lot number and manufacturing time. The content of the database is protected from unauthorized access and is only accessible to a known set of reading devices.

Product checks. Inspections are usually conducted in storehouses and on shop floors. For the product category it is desirable that individual items do not have to be unpacked, and tests in shops need to be performed in an unobtrusive manner so that potential customers are not disturbed. Employees or contractors are equipped with handheld reader devices for the inspection process. Individual readers are registered with the company to restrict unauthorized access. The network address of the company's service is known to the devices, so no address lookup service is required. In order to determine whether the serial number and the Tag ID are valid, an inspector usually has to pick up a package, hold the reader device in the proximity of the package (for example less than 10 cm to 5 cm away from a defined spot), and wait for a few seconds for the device to query the company's database via a GSM or GPRS modem. The status of the test is shown on a simple display. Only if the test indicates irregularities does the package have to be opened for a physical inspection of for example the imprinted serial number or other covert security features.

Advantages. The insular solution is relatively easy to set up. It does not require collaboration among numerous stakeholders; the potentially difficult process of finding a compromise with respect to data sharing, access rights, etc. is not a prerequisite. Although using standardized components could save transponder costs and allow for easier integration with other systems at a later point in time, the limited complexity of the insular solution also allows for choosing proprietary tags and readers. This may be desirable if additional features should be integrated such as sophisticated authentication protocols or warranty receipts on tags.

Hurdles. A financial analysis of the non-standard-based insular approach has provided strong evidence that low costs for system setup are counterbalanced by the time-consuming authentication process. Relying on a field force that visits individual retail stores is unlikely to result in interception rates that justify the investment for the product category under study. Furthermore, the low expected interception rates would neither lead to indirect effects that could limit counterfeit supply nor noticeably improve insights into the counterfeit market.

Application scenario II: Using RFID to support customs

The second use case outlines the potential application of RFID to support customs during the inspection process. For most countries with strictly enforced intellectual property rights counterfeit imports constitute the major source of illicit imitation products. Since customs can be regarded as the only organization which has authority over imports on a regular basis when goods cross an external border, customs is a powerful stakeholder in the anti-counterfeiting battle. A technology which helps to authenticate incoming goods can speed up the inspection of those consignments that are equipped with the technology, and thereby provide officers with more time to check other goods. The basic idea behind the approach is to allow customs to conveniently determine if a consignment is being shipped from and to a trusted party, and if the declared goods are the goods which are being shipped. A principle use case for a "trusted shipping" approach is briefly outlined below.

Production and shipment. In order to participate in a trusted shipping program, the dispatcher has to register with customs and provide them with the network address of their service access point.[57] A certification process as well as authentication of the provided network address is crucial in order to prevent illicit actors from feeding bogus references into the system. When products are prepared for shipment, the dispatcher generates an entry in a database system which contains:

- a unique consignment identifier,
- a description of the products and the quantities shipped, including the serial numbers of the individual items (if available),
- information on the source and destination with company identifiers (for example a tax registration number) and postal addresses,
- the weight of the consignment,
- the address of an online resource for further counterfeit-related information (for example a description of the security features or distinguishing characteristics of counterfeits), and/or contact information for further enquiries,
- a time stamp with the associated maximum "age" (or time-to-live) for the data entry, and
- other information that customs officials may need in order to process the shipment declaration.

---

[57] Please note that, unlike the previous approach, a suitable customs system does not yet exist for small consignments, though it would be technically feasible to implement one.

Furthermore, the dispatcher attaches an RFID transponder to the consignment and writes its company identifier, the consignment identifier, and a password that allows customs to access the data entry associated with the consignment during the defined time-to-live of the entry on the responder. The Tag ID is appended to the database entry for additional security. UHF transponders are used to facilitate larger read ranges (to the order of 5 meters); memory capacities of about 16 bytes on the tag are sufficient. In order to ease the inspection process, the consignment is labeled with a sign helping the inspection personnel to recognize and find the transponder.

Product checks. Products that arrive at an external border are to be declared. During the declaration customs officials recognize those consignments which facilitate electronic references to trusted shipping partners from the shipments' labels. Transponders are read either by a handheld device or by a fixed reader when the consignment passes a dock door. Using the consignment identifier as an input, the reader (or the user terminal) retrieves the network address of the dispatcher's online service. With the consignment identifier and the temporary password, the database is accessed to obtain the corresponding information. The weight of the consignment may further be determined to compare it with the corresponding entry. If the product check did not reveal any irregularities, the freight papers are printed out, possibly already including the information retrieved from the company's database and extended by the officers' data entries. If designed carefully, the approach could significantly accelerate the inspection process and allow for differentiating inspection rates, leaving more time to check suspicious goods. In order to motivate companies to participate in a trusted shipping program, customs may grant preferential treatment to consignments that are part of such an agreement.

Given the enormous variety and volume of products crossing a border each day, a standardized solution is a prerequisite; it is, for example, not acceptable for customs to operate several inspection devices. However, the relationship between customs and dispatchers resembles a 1:m (one to many) structure, for which access management and data sharing is easier to establish as in n:m (many to many) relationships. Furthermore, data is not shared among potential competitors or organizations which may aim to integrate process steps or use a higher supply chain visibility to strengthening their position towards the company providing this information. In this way no direct threats are posed to the participating enterprises.

Advantages. An application of RFID at case level to support customs is in fact very promising. Relying on a standardized solution that allows for a seamless integration of product authentication in an existing inspection process makes high sample rates possible. Due to the use of standard reader devices, their costs can be shared among the participating companies. Moreover, seizures at customs often concern large quantities (since the products arrive in bulk) and can reveal valuable information on the shipment tactics or even on other illicit actors. Additional potential benefits include faster declarations for licit goods as the system could effectively support the paperwork.

Hurdles. The approach requires an agreement upon the standards for transponder, reader, data exchange and processes to govern the inspections and exception handling. Although the EPC Network is well suited as a technical basis, the different countries and economic areas still have to outline how a common system should look and who would operate it.

Application scenario III: Using a track-and-trace system

The third application scenario builds upon the emerging EPC Network which can help to establish plausibility checks for a product's authenticity. In theory the infrastructure allows individual companies, customs, supply chain partners and even end users to retrieve information on the origin and history of an article. Given the numerous application scenarios and the preliminary status of the underlying infrastructure, the following description is more generic than the previous ones.

Production and information collection. An organization (referred to as EPC Manager), typically at an early step of the supply chain, assigns a unique identifier (the EPC) to a product, stores it on an RFID transponder, and generates a database entry with information – for example on the product type and its specification, time of production, etc. – that can be referenced over the corresponding EPC. Supply chain events such as a change of location or ownership are recorded and linked to the respective database entry. Subsequent owners also generate data entries in their system in order to record date of purchase, previous owner, aggregation, usage as components in other products, location of storage, etc. An alternative solution to decentralized data storage could be based on extending and passing on individual data records to the next owner of an object once the location, status or ownership of the item has changed; in this scenario, a physical object is accompanied by the corresponding data.

Product checks. To conduct a check, a user (i.e. a customs officer, a supply chain partner or a consumer) would scan the transponder and thereby initiate a search or lookup within the EPC Network. A network service can, given an EPC number, return address information of the corresponding database entry. If the requesting party has the necessary access rights, this would allow the product's history to be assembled. Heuristics can be applied to assess whether the retrieved track-and-trace information is plausible.

Benefits. The use of the EPC Network to compile track-and-trace data allows for almost complete monitoring of the supply chain. Product checks can, for example, be arranged at the goods receiving department or the point of sales, and would deter counterfeit producers from selling deceptive counterfeit articles. The solution could easily support customs with the implications that we outlined before. Track-and-trace systems offer many other benefits, for example with respect to distribution control, direct selling, quality assurance, etc.

Hurdles. As we already outlined in Section 9.3, the implementation of cross-industry track-and-trace systems is quite challenging. Hurdles include objection to data sharing among different companies, problems with access rights management, the need for comprehensive standards, conflicting interests among different industries on how the system should be implemented, and the enormous amount of data such systems generate. Moreover, though the costs per product authentication will be low once the system is implemented, the set-up costs are high. They are likely to be justified only if non-counterfeit-trade-related benefits are considered as well, which additionally complicates adoption decisions.

Implications for monitoring, reaction, and prevention activities

RFID-based anti-counterfeiting techniques allow for frequent product inspections throughout the supply chain. For monitoring activities the benefits are rather obvious. With RFID it becomes feasible to integrate security checks into existing processes, for example at the registration of incoming goods or at a retailer's check-out. Warehouse inspections can be performed at low cost by untrained staff rather than by anti-counterfeiting experts. The technology ultimately leads to higher inspection rates and helps to overcome the shortcomings of many established technologies which are inexpensive to apply, but require time-intense inspection procedures. User-friendly interfaces, machine-assisted inspections and adequate protection against tag cloning further increase the reliability of the tests, which in turn increases the interception rates of counterfeit goods.

However, RFID can alter monitoring activities not only with respect to speed, cost and reliability. Many of the above-mentioned benefits stem from the potential to integrate external stakeholders in the authentication process. Therefore monitoring approaches are likely to evolve from an activity with a focus on market investigations to stakeholder management where companies try to get others knowingly or unknowingly to support their anti-counterfeiting measures. Brand-protection managers have to decide where automated inspections are most efficient (i.e. where the average cost to identify a counterfeit article is low), where existing processes allow for a seamless integration of reader devices (for example on an assembly line), where other suppliers or customers have an interest to set up and share a respective infrastructure, how brand owners can work together to support customs or convince other stakeholders to adopt the solution, etc. Furthermore, in cases where inspections require user interaction, it is worth considering how product authentications can be positioned as a value-adding service rather than as additional labor input. In this context brand owners may also consider trying to involve consumers in product-protection measures. This can be promising if imitations are perceived as a substantial risk by the consumer, or if checks can be implemented as implicit steps of other interactions. These may include becoming part of sweepstakes, mail-in rebates, discount or loyalty systems, etc.

Regardless of where product inspections take place, improved monitoring activities lead to a larger number of seizures and thus to more frequent subsequent actions. The associated learning effects help to enhance enforcement strategies and facilitate prosecution and seizure efforts. Most reaction-related improvements, however, only indirectly relate to the application of RFID. As the technology can assist external stakeholders, it may help to strengthen their collaboration with licit manufacturers. Failing product authentications could, for example, automatically trigger further steps without requiring the third party to initiate contact with the right holder. Resources which become available due to more efficient monitoring processes can be used to strengthen reaction measures. The overall cost of reaction efforts is nevertheless likely to increase given a larger number of cases that become known and the effort associated with dealing with them. However, if a company does not want to follow a zero-tolerance strategy, it becomes possible to select the more important cases on a wider basis, which can in turn increase the efficiency of the reaction steps. The causal directions are illustrated in Figure 9.5.

Preventive measures include steps to limit the risk that originates from existing substandard imitation products as well as steps to confine future counterfeit supply. Protection against existing imitations aims to identify such goods or, put another way, to ensure the authenticity of the items that are to be used. For articles that are in the possession of a company, RFID makes even complete checks feasible. An advantage of the approach is that the corresponding processes resemble internal monitoring activities. Future counterfeit supply is affected by changing the risk-return consideration of illicit actors. The implications for the different types of counterfeit producers are outlined below.

Implications of RFID for the different types of counterfeit producers

Counterfeit producers appear to base their operations on concisely-defined production settings and rely on cooperation with various other illicit stakeholders. As outlined before, they resemble licit enterprises as they at least implicitly perform investment-risk-return considerations and are likely to only engage in a venture if the corresponding business case is promising and more attractive than alternative licit or illicit activities – such as counterfeiting products of less-protected brands. RFID technology can alter the underlying cost-benefit considerations directly as mimicking RFID-based features is prohibitively expensive, and indirectly, as large-scale checks help to confiscate products without such mechanisms. The extent of the effects, however, depends on the individual strategic settings of illicit actors (c.f. Section 2.1 and Section 2.3). Their different approaches to risk and the varying

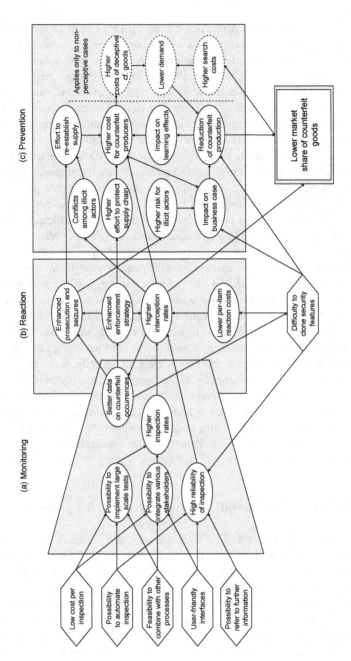

Figure 9.5: The impact of RFID on (a) monitoring, (b) reaction, and (c) prevention

cost of production that each strategic setting exhibits consequently also affect the implications of product seizures and thus the effectiveness of anti-counterfeiting technologies.

Disaggregators mostly concentrate on inter-personal brand-related free-rider effects and sell their goods as non-deceptive counterfeits. For such goods we found consumers to be particularly sensitive to product price and search cost. The latter are both strongly affected by seizures, as the number of confiscated products goes up, the price of the counterfeits will rise since the expenses for the seized articles have to be covered. The effect is more pronounced for articles which have a high variable production cost. Moreover, larger-scale operations rely on middle men in different countries, and consequently on business relationships that, given the clandestine nature of the market, are time and cost-intensive to establish. Intermediate stakeholders will claim higher margins due to the increased risk, which again influences sales prices of imitation products. This will lead to less supply, and finding non-deceptive counterfeits will become more difficult.

Imitators typically have high production costs relative to the other types of counterfeit producers. They manufacture products with a relatively high functional level of quality. Seizures are especially painful for this group as they lead to high losses and can endanger expensive production facilities. For this reason, Imitators sell their goods mostly to their home markets where the risk imposed by legal steps from the brand owner is typically low. Since they avoid countries where intellectual property rights are strictly enforced and their domestic market is often sufficient alone, protection technologies are not likely to have a strong effect on future counterfeit production. However, high quality counterfeits are sometimes shipped into other countries by illicit actors that are not associated to their producers, and authentication technologies can prevent others from unintentionally using or consuming them.

Fraudsters and especially *Desperados* often take high seizure rates into account. They are, due to their limited investments, typically not prone to the confiscation of production equipment. Though higher interception rates make the supply of such goods more difficult, enforcement strategies remain hard to implement as the actors typically operate only for a short period of time, do not require an expensive infrastructure, and may choose to focus on other illicit activities shortly after a number of counterfeits have been shipped. As their primary sales channel at least in Europe and North America is the Internet, Fraudsters and Desperados also bypass many product checks, for example at retail markets. However, with respect to the targeted products, consumers have a strong interest in purchasing genuine goods and are easily motivated to verify the authenticity of potentially dangerous articles; RFID enabled mobile phones could facilitate such inspections, but their adoption is still to take place.

Counterfeit Smugglers realize large profits not only due to brand-related free-rider effects, but also by evading taxes. The tobacco market constitutes a good example. Here most shipments arrive in sea freight containers which can hold as many as 10 million cigarettes with a market value of more than EUR 2 million. In the European Union and the United States taxes of well above 100% would justify the direct costs due to seizures of every second consignment; consequently even a significant increase of seizure rates is unlikely to directly alter the underlying business case. However, higher interception rates can nevertheless reduce the counterfeiters' margins, lead to better chances of successful prosecutions, and significantly increase the risk for illicit actors. In fact, given the high demand for counterfeit cigarettes, the existing countermeasures already appear to be the major limiting factor of trade with such goods, and RFID can effectively support these measures.

For all types of counterfeit producers the direct losses due to product seizures only account for one of the implications of efficient product identification measures. Side effects – such as the increased risk for the actors and the higher margins seized and the cost to reorganize the distribution channel once it has been compromised – affect their overall business cases and play a very important role in preventing future counterfeit market activities. In fact, seizures lead to a reduction of counterfeit sales that goes far beyond the actual number of confiscated articles. Their preventive character should be taken into account when considering investment in security technologies.

Part E   Managerial Guidelines and Conclusions

10 Guidelines

With this book we have sought to improve practical understanding of the counterfeit market. We have introduced a set of tools to determine the volume of illicit trade, discussed the implications for brand owners and licit manufacturers, and described how successful companies respond to trademark infringements. Moreover, we have identified a set of generic strategies that illicit producers tend to follow and outlined the underlying business rational as well as the corresponding investment-risk-return considerations. The supply-side investigations revealed specific strengths and weaknesses with respect to susceptibility to seizures, access to capital and labor, prospects of future growth and risks for those who run and finance the illicit venture. Together with insights into the shipment and distribution tactics and a good understanding of consumer demand and awareness, it becomes possible to develop brand-, product-, and market-specific anti-counterfeiting strategies that can effectively reduce trademark infringements. The following guidelines highlight what brand owners should do to protect their assets and safeguard their customers.

1. Determine the market share of counterfeit goods

The penetration of counterfeit goods is highly dependent on the product category, brand and geographic market. Even industry-specific estimates rarely adequately reflect the situation for individual companies. Reliable market studies, however, constitute the basis for further risk and impact analyses. They enable companies to better allocate resources to specific countries or product groups. Therefore management should start the refinement of their anti-counterfeiting program with a solid analysis of the penetration of illicit imitation products. The computational framework that has been introduced in Chapter 6 is well suited for such analyses. The calculation should be performed for each key brand and product as well as for different geographic markets. The results help to set an initial focus for further investigations.

2. Investigate the characteristics of the counterfeit producers

Competitor analyses are an indispensable tool for the development of corporate strategy. In the same way supply-side analyses of the illicit market can help companies to develop effective strategies to respond to counterfeit actors. An investigation of the production settings and strategies often reveals specific strengths and weaknesses of illicit actors and allows for a better prediction of their future behavior. Like competitor analyses, such considerations help to prioritize reactive and preventive measures as they enable the brand owner to concentrate efforts on the opponent's weak spots. Therefore, in step 2, we recommend revisiting Section 2.1 to ascertain the strategic settings that are dominant for the brands and products under consideration. Moreover, we recommend outlining potential business plans from a counterfeiter's perspective and try to figure out what factors limit the speed of growth of the illicit venture. These limiting factors are a good indicator of where companies can bring leverage to bear on counterfeit producers.

3. Understand the properties of the illicit supply chain

Illicit actors can make or buy inputs, transfer outputs downstream, or sell them. In fact, counterfeit goods exist in final and intermediate markets. In order to protect licit companies from imitations infiltrating their supply, managers have to eliminate the value chain's permeability to such goods. Moreover, to increase interception rates among articles that are traded in parallel to the licit supply chain, traffic routes and shipment patterns have to be investigated. This, however, requires some knowledge of the structure of the illicit market and the way it interfaces with the licit supply chain. The market model that has been introduced in Section 2.2 helps to systematically collect the required information. Prior to the definition of supply chain security measures, one should instantiate a model for each brand and product under study. Even if counterfeit goods mostly appear in the final market, brand owners should start their analyses way down in the value chain, maybe even with the raw materials and components that the producers require. Typical import routes, batch sizes, shipment strategies, and entry points in the licit supply chain should be highlighted. Companies that fear that substandard imitations can infiltrate their own parts supply should involve their sub-contractors when an illicit supply chain model is instantiated. After step 3 and together with the insights into the production settings of illicit actors, managers are already able to evaluate how many different actors are involved in producing and selling the imitations and whether manufacturing or distributing them is the major challenge for their opponents.

4. Analyze the behavior of counterfeit consumers

Consumers may have a strong interest in purchasing genuine items, but may also invest considerable effort in finding cheap imitation products. Consequently their behavior directly affects the level of support that brand owners can expect from potential customers when it comes to identifying counterfeit articles. Knowledge of consumers' awareness of counterfeit products, the willingness to buy such goods, and their reasoning for or against purchasing is fundamental to consumer education efforts. This knowledge also provides the basis for substantiated impact analyses. Brand owners should also learn what people think or know with respect to trademark infringements. Since attitudes and awareness are highly dependent on the individual product and brand, it is important to gain a thorough understanding with respect to the specific goods under consideration. Especially for product categories with a high share of counterfeits in the market, we recommend conducting at least one pretest at an early stage of the anti-counterfeiting strategy development. Section 3.2 describes how suitable market investigations can be set up.

5. Conduct a risk analysis and assess the monetary loss

Based on the market share and demand analyses, the risks and costs associated with counterfeit trade can be evaluated. Such considerations should include a qualitative assessment of the risk of additional liability claims and the impact on future competition (especially in emerging markets), as well as quantitative analyses of the loss of revenue and the impact on brand value. The set of tools and evaluation guidelines introduced in Chapter 6 can ease these investigations. We are aware that many practitioners reject the idea of quantitative impact analyses; they rightly say that steps against potentially dangerous goods should be taken in any case, and monetary analyses are inappropriate when the consumers' health and safety is at risk. However, the financial means of companies are limited, and management has to allocate the funds to different products and geographic markets. Knowledge of the financial loss due to illicit imitation products forms the basis of a substantiated prioritization of the problem, and makes business case calculations on countermeasures possible. If quantitative assessments are not feasible, management should at least commission a thorough scenario-based risk analysis.

6. Analyze best practice strategies

Some companies already have successful brand- and product-protection strategies in place. Rather than developing everything from scratch, it is worth identifying best practice approaches and seeing whether they can, at least in part, be adopted. Management should also consider anti-counterfeiting measures from different industries that often have a different but complementary perspective on the illicit market. The benchmarking study presented in Chapter 4 may serve as a starting point for step six as it details the characteristics of successful monitoring, reaction, and prevention efforts.

7. Set up or refine your brand- and product-protection task force

After steps one to six, one should have a good idea about the scale of the problem as well as of the products and markets that are the most affected. Moreover, the basic characteristic of successful anti-counterfeiting measures should be known. With this background information the brand- and product-protection task force can be set up. Step seven is probably the most difficult one; the organizational structure should facilitate a lively information exchange throughout the company, signal the importance of the topic, allow for centralized decision making and at the same time build upon local contacts with external stakeholders; the distribution of cost among different business units has to be defined, and basic target agreements for the involved employees must be drafted. We have discussed the properties of suitable organizational structures in Section 5.4.

8. Implement defined monitoring and reaction processes

The initial market share, risk and impact analyses help prioritize the problem and to identify medium and long-term trends within the counterfeit market. They should be repeated on an annual basis and serve as an input when the overall anti-counterfeiting strategy is revised. On an operational level, however, market investigations must enable timely response to signs of counterfeiting. Companies have to have continuous monitoring processes in place and at the same time make sure that tip-offs from supply chain partners, customs, competitors and end users are followed up. In fact, effective monitoring activities are the key element of successful anti-counterfeiting efforts. Management has to make sure that awareness of the problem and the commitment to fight counterfeiting is present throughout the company. The actual design of monitoring processes, however, requires deep insights into the supply- and demand-side characteristics of the illicit market; Section 5.1 describes the most important properties of such activities. The configuration of the individual monitoring tasks is the responsibility of the brand- and product-protection task force. However, management should encourage the development of performance measures especially with respect to market observations.

Reaction usually consists of withdrawing counterfeit articles from circulation, seeking to prosecute offenders, managing the relationship with informants and affected individuals or companies, and refining the anti-counterfeiting strategy. Consequently the reaction to occurrences of counterfeiting not only limits potential damage, but also has a strong impact on the performance of future anti-counterfeiting measures. The individual properties of the reaction processes are highly dependent on the scope of the problem; their principle design is discussed in Section 5.2. Management should specify how much effort should be invested to respond to non-critical counterfeit cases, call for the development of contingency plans that facilitate quick responses, and foster learning effects from both successful and unsuccessful reactions.

9. Assess and select preventive measures

Preventive measures typically aim to secure the company's supply chain, eliminate production of counterfeit goods, hamper their distribution and discourage or prevent users or consumers from purchasing faked goods. The means to reach these goals can be organizational, technological, legal, and communicative in nature. As the benchmarking study has revealed, successful companies use a combination of all of the above approaches but especially focus on awareness raising and lobbying for better enforcement of intellectual property rights. Companies that particularly fear the emergence of counterfeit articles in their own parts supply stress the importance of awareness raising within purchasing, quality management and production. They should also aim to strengthen the commitment of their subcontractors and component suppliers. For product categories where counterfeit goods are primarily sold into final markets, the preventive measures which seem to be most promising are those that increase the cost and risks for counterfeit actors and raise the search cost for counterfeit consumers. When choosing selective measures, brand owners should again ask themselves what factors limit the speed of growth of counterfeit activities and consider steps that strengthen the influence of these factors. In general investment-risk-return considerations of counterfeit producers appear to strongly influence future supply. Therefore management should regard attempts to seize illicit products and prosecute offenders not only as a part of the reaction processes but also as a cornerstone of prevention.

10. Consider the implementation of large-scale product checks

High seizure rates have been shown to have a considerable impact on counterfeit supply. For illicit actors they cause losses far beyond the costs of the intercepted goods as they also increase the risk of prosecution, may require the development of new shipment routes with new intermediate stakeholders, and constitute causes for conflicts among those who were involved in the failed delivery attempt. Product security features can help to authenticate genuine products and conversely to identify imitations and realize high interception rates. However, most established security features may be highly resistant to cloning attacks but appear to be not suited for large-scale product checks by untrained staff. Consequently counterfeit producers can rely on low inspection rates and do well with poor imitations of features that would be extremely difficult to precisely duplicate. The use of covert features may play some role in fending off liability claims after an incident has occurred, but is unlikely to significantly reduce the number of illicit products within the supply chain. Therefore we clearly recommend focusing on overt, easy-to-inspect features. Moreover, when selecting a technology, brand-protection managers should consider the price per feature and the cost and effort per inspection.

Again, if inspections cannot be conducted by a large number of people in a convenient way, they are only of little use. RFID technology helps to overcome several shortcomings of established anti-counterfeiting technologies (c.f. Chapter 1). If carefully designed, RFID transponders can effectively avert cloning attacks, provide a migration path towards a higher level of security if needed, and – what is most important – require limited effort and expertise during inspection. Though individual transponders are relatively expensive compared to established features such as holograms, they have been found to be cost-effective in process settings that facilitate frequent checks. Consequently they are well suited when it is possible to integrate product inspections seamlessly into existing processes – as shown for an application to support clearances at customs – but are less likely to be superior to other technologies if such checks have high overhead costs – for example when relying on dedicated inspection teams that have to travel to the point of sale or a warehouse. In any case, we recommend considering the use of automated identification technologies that increase supply chain visibility and thereby help to protect against counterfeit goods.

11. Integrate external stakeholders in your anti-counterfeiting efforts

Brand owners are not the only actors who have a strong interest in the integrity of supply. With respect to own sourcing activities, the problem awareness of component suppliers and their willingness to pass on the product security efforts further downstream in the value chain should become part of the purchasing decision. For upstream stakeholders including wholesalers, retailers, customs and end users, it may be easier and thus more cost-efficient to authenticate products than for the brand owner. When defining the company's anti-counterfeiting strategy, management should outline whom to integrate in such efforts, and what incentives to provide to those who participate. For commercial partners such incentives may be some sort of cost sharing or the possibility to use the identification technology for other purposes as well. With RFID such additional benefits could be automated-goods-receipt or check-out processes. Ideally product authentication can be positioned as a service so that the added value becomes apparent for the consumer.

12. Signal top-management support

Anti-counterfeiting measures may not immediately have a positive business case, and for many employees it is difficult to see the benefit of the efforts at all. Even worse, counterfeit incidents are often perceived as a mishap that somebody else is responsible for and consequently somebody else should take care of. Support from top management is crucial to signal that the problem has to be proactively addressed and that attentive employees make an important contribution to the wealth of the company.

13. Revisit the brand- and product-protection effort regularly

Counterfeit producers try to get as much out of their market as possible. They will continue to explore new business concepts, target more sophisticated or less protected product categories, and utilize more advanced production techniques. Consequently the brand- and product-protection measures should be revisited periodically and have to be refined if changes in the counterfeit market become apparent.

11 Concluding Remarks

Counterfeit trade on the current scale constitutes a relatively new but significant challenge for many industries. Licit manufacturers and brand owners face numerous forms of counterfeit production, threats to their supply and distribution channels, ambiguous consumer behavior, and partially insufficient legal protection. Successful anti-counterfeiting measures require a systematic approach that reflects the complexity of the phenomenon. However, many companies have yet to build up the know-how to combine organizational, legal, and technical measures into effective monitoring, reaction, and prevention strategies. Such strategies should be based on a thorough understanding of the characteristics and production settings of counterfeit producers, knowledge of the demand-side and sales channels, as well as some familiarity with the flow of goods and the role of individual stakeholders.

Appropriate measures appear to be by no means without effect. Companies with anti-counterfeiting strategies that are founded in a good understanding of the market mechanisms of counterfeit trade, that are strongly supported by senior management, that are based on well-defined monitoring and reaction processes, and that utilize strong collaboration with external stakeholders rarely see the need to adjust their market and brand positioning as a result of counterfeit trade. On the other hand, companies that have no adequate measures in place are much more likely to consider drastic, potentially expensive steps, including a withdrawal from certain markets – which is likely to reduce growth and long-term profit.

Moreover, high seizure rates have the potential to severely change the investment-risk-return considerations of counterfeit producers – to the point where an illicit actor's business case to reproduce and sell protected products becomes less promising than engaging in other illegal or legal activities. In this respect, RFID-based anti-counterfeiting technologies constitute a promising approach to disrupt the flow of counterfeit goods and to protect the integrity of the supply chain.

Another important factor that determines future production of counterfeit goods is the development of the intellectual property landscape in Asia's emerging economies. Before joining the World Trade Organization in 2001 China, for example, strengthened its legal framework to comply with the Agreement on Trade-Related Aspect of Intellectual Property Rights. However, despite stronger statutory protection and sporadic, drastic measures against producers of hazardous goods, foreign right holders still face major problems when trying to convince Chinese courts to take actions against counterfeit actors. This especially applies to goods where the consumers' health and safety is not directly at stake, when counterfeit producers are important regional employees, and when their output is of some value for the domestic market. In fact, many of those producers can be expected to develop into more or less licit competitors once intellectual property rights are more strictly enforced. The longer they have time to utilize the free-rider effect, the longer they learn how to manufacture and sell goods. Therefore, and despite high hopes in new supply chain security measures, brand owners must continue to lobby for a more rigorous enforcement of intellectual property rights. In newly industrializing countries the situation is unlikely to change unless their governments realize that they become on balance a victim instead of a benefactor of counterfeiting. Until then brand owners and licit manufacturers have to diligently protect their own supply, limit damage to their brands, and persuade potential consumers that only genuine goods convey the values a trademark stands for. Successful companies might even be able to leverage the increased brand awareness and use the opportunity to strengthen their supply chain visibility and distribution control.

We hope that our book will help you to gain the market insights that are necessary to protect consumers and safeguard the assets of companies. It should pave the way towards a substantiated and fact-based managerial response that better reflects the complex and dynamic properties of the illicit market.

Appendix

List of References

Aaker DA (1996a) Managing brand equity. New York, NY: The Free Press.

Aaker DA (1996b) Measuring brand equity across products and markets. California Management Review, 38(3): 102–121.

ACG – Anti-Counterfeiting Group (2003) Why you should care about counterfeiting? High Wycombe, United Kingdom: Anti-Counterfeiting Group.

Albers-Miller ND (1999) Consumer misbehavior: Why people buy illicit goods. Journal of Consumer Marketing, 16(3): 273–287.

Albrecht K and McIntyre L (2005) Spychips: How major corporations and government plan to track your every move with RFID. Nashville, TN: Nelson Current.

Aldenderfer MS and Blashfield RK (1984) Cluster analysis. Newbury Park, CA: Sage.

Ang SH, Cheng PS, Lin EAC, and Tambyah SK (2001) Spot the difference: Consumer responses towards counterfeits. Journal of Consumer Marketing, 18(3): 219–235.

Bach D (2004) The double punch of law and technology: Fighting music piracy or re-making copyright in a digital age? Business and Politics, 6(2): 1–35.

Balkin DB, Shepherd DA, and De Castro JO (2004) Piracy as strategy? A reexamination of product piracy. Working Paper Economia wp04-08, Instituto de Empresa.

Bamossy G and Scammon L (1985) Product counterfeiting: Consumers and manufacturers beware. Advances in Consumer Research, 12(1): 334–339.

Barnett J (2005) Shopping for Gucci on Canal Street: Reflections on status consumption, intellectual property, and the incentive thesis. Virginia Law Review, 9: 1381–1423.

BBC News (2005) China to tackle software piracy. Updated April 12, 2006. http://news.bbc.co.uk/hi/technology/4902976.stm.

Ben-Shahar D and Assaf J (2004) Selective enforcement of copyright as an optimal monopolistic behavior. Contributions to Economic Analysis & Policy, 3(1): Article 18.

Bettman JR, Luce MF, and Payne JW (1998) Constructive consumer choice processes. Journal of Consumer Research, 25(3): 187–217.

Bian X and Veloutsou C (2007) Consumers' attitudes regarding non-deceptive counterfeit brands in the UK and China. The Journal of Brand Management, 14(3): 211–222.

Bloch PH, Bush RF, and Campbell L (1993) Consumer 'accomplices' in product counterfeiting. Journal of Consumer Marketing, 10(4): 27–33.

Bourn J (2005) Comptroller and auditor general's standard report on the accounts of HM customs and excise 2004-05. National Audit Office, London.

Braun OL and Wicklund RA (1989) Psychological antecedents of conspicuous consumption. Journal of Economic Psychology, 10(2): 161–187.

Bryce J and Rutter J (2005) Fake nation? A study into an everyday crime. Belfast, Northern Ireland: Organised Crime Task Force. http://les1.man.ac.uk/cric/Jason_Rutter/ppers/Fake Nation.pdf.

BSA – Business Software Alliance (2006) Third annual BSA and IDC global software piracy study. Washington, DC: Business Software Alliance.

Bundesministerium der Finanzen (2002) Die Bundeszollverwaltung – Jahresstatistik 2001. Berlin: Bundesministerium der Finanzen.

Bundesministerium der Finanzen (2006) Die Bundeszollverwaltung – Jahresstatistik 2005. Berlin: Bundesministerium der Finanzen.

Bush RF, Bloch PH, and Dawson S (1989) Remedies for product counterfeiting. Business Horizons, 32(1): 59–65.

Business Week (2005) Fakes! February 2: Cover story. www.businessweek.com/magazine/content/05_06/b3919001_mz001.htm.

Cas J (2005) Privacy in pervasive computing environments – A contradiction in terms? IEEE Technology and Society Magazine 24(1): 24–33.

CEBR – Centre for Economics and Business Research (2002) Counting counterfeits: Defining a method to collect, analyse and compare data on counterfeiting and piracy in the single market. Centre for Economics and Business Research, London, United Kingdom.

Chakraborty G, Allred AT, and Bristol T (1996) Exploring consumers' evaluations of counterfeits: The roles of country of origin and ethnocentrism. Advances in Consumer Research, 23: 379–384.

Chakraborty G, Allred A, Sukhdial AS, and Bristol T (1997) Use of negative cues to reduce demand for counterfeit products. Advances in Consumer Research, 24(1): 345–349.

Chang MK (1998) Predicting unethical behavior: A comparison of the theory of reasoned action and the theory of planned behavior. Journal of Business Ethics, 17(16): 1825–1834.

Chaudhry PE and Walsh MG (1996) An assessment of the impact of counterfeiting in international markets: The paradox persists. Columbia Journal of World Business, 31(3): 34–48.

Chaudhry PE, Cordell V, and Zimmerman A (2005) Modeling anti-counterfeiting strategies in response to protecting intellectual property rights in a global environment. The Marketing Review, 5(1): 59–72.

Chaudhry PE (2006) Managing intellectual property rights: Government tactics to curtail counterfeit trade. European Business Law Review, 17(4): 939–958.

Cheung W-L and Prendergast G (2006) Buyers' perceptions of pirated products in China. Marketing Intelligence & Planning, 24(5): 446–462.

Chiou J-S, Huang C-Y, and Lee H-H (2005) The antecedents of music piracy attitudes and intentions. Journal of Business Ethics, 57(2): 161–174.

Chuchinprakarn S (2003) Consumption of counterfeit goods in Thailand: Who are the patrons? European Advances in Consumer Research, 6: 48–53.

Clark A (1997) Enforcement of intellectual property rights. Working Paper, University of Warwick, United Kingdom.

Clark D (2006) Counterfeiting in China: A blueprint for change. The China Business Review, January/February: 14–15 & 46–49.

Cole PH and Ranasinghe DC (2008) Networked RFID systems and lightweight cryptography – Raising barriers to product counterfeiting. Berlin, Germany: Springer.

Conlisk J (1996) Why bounded rationality? Journal of Economic Literature, 34(2): 669–700.

Conner KR and Rumelt RP (1991) Software piracy: An analysis of protection strategies. Management Science, 37(2): 125–139.

Cordell VV, Wongtada N, and Kieschnick RL (1996) Counterfeit purchase intentions: Role of lawfulness attitudes and product traits as determinants. Journal of Business Research, 35(1): 41–53.

Culnan MJ and Armstrong PK (1999) Information privacy concerns, procedural fairness, and impersonal trust: An empirical investigation. Organization Science 10(1): 104–116.

Culnan MJ and Bies RJ (2003) Consumer privacy: Balancing economic and justice considerations. Journal of Social Issues 59(2): 323–342.

Das R and Harrop P (2008) RFID forecasts, players & opportunities 2008-2018. IDTechEx, Cambridge, UK. www.idtechex.com/products/en/view.asp?productcategoryid=151.

De Castro JO, Balkin DB, and Shepherd DA (2006) Can entrepreneurial firms benefit from product piracy? Journal of Business Venturing, to appear.

De Matos CA, Ituassu CT, and Rossi CAV (2007) Consumer attitudes toward counterfeits: A review and extension. Journal of Consumer Marketing, 24(1): 36–47.

Die Zeit (2006) Der Pirat wird Erfinder. August 25. No. 35. www.zeit.de/2006/35/Produktpiraterie_China.

Dubois L (2006) The situation of counterfeiting in Japan. Talk given at the EU-Japan Business Dialogue Round Table, Tokyo, Japan, July 13–14. www.eujapan.com/roundtable/presentation_dubois2_06.pdf.

EC – Commission of the European Communities (2005a) Communication from the commission to the council, the European parliament and economic and social committee on a customs response to latest trends in counterfeiting and piracy. Document ID: COM(2005) 479 final. Brussels, Belgium: European Commission.

EC – European Commission Communication (2005b) Commission launches action plan to combat counterfeiting and piracy. Press release, October 11. Document ID: IP/05/1247. http://europa.eu/rapid/pressReleasesAction.do?reference=IP/05/1247&format=HTML&aged =0&language=EN&guiLanguage=EN.

EC – EU/US Summit (2006) EU/US action strategy for the enforcement of intellectual property rights. Meeting protocol, June 21, Vienna, Austria. www.eurunion.org/partner/summit/ 20060621sum.htm.

EC – European Commission Taxation and Customs Union (2007) Summary of the community of customs activities on counterfeit and piracy results at the European border 2006. http://ec.europa.eu/taxation_customs/resources/documents/customs/customs_controls/counter feit_piracy/statistics/counterf_comm_2006_en.pdf.

EC – European Commission Taxation and Customs Union (2008) How can right holders protect themselves from counterfeiting and piracy? http://ec.europa.eu/taxation_customs/customs/ customs_controls/counterfeit_piracy/right_holders/index_en.htm.

ECR Europe (2003) Optimal shelf availability: Increasing shopper satisfaction at the moment of truth. ECR Europe, Brussels, Belgium, 2003.

ecma international (2005) Near Field Communication. www.ecma-international.org/activities/ Communications/tc32-tg19-2005-012.pdf.

Eckfeldt B (2005) What does RFID do for the consumer? Communications of the ACM 48(9): 77–79.

Eisend M and Schuchert-Güler P. (2006) Explaining counterfeit purchases: A review and preview. Academy of Marketing Science Review, 10(12): 1–25.

EPC ARC – EPCglobal (2005) The EPCglobal architecture framework. Final version, July 1. www.epcglobalinc.org/standards/Final-epcglobal-arch-20050701.pdf.

EPC ONS – EPCglobal (2005) Object naming service (ONS). EPCglobal ratified specification version 1.0, October 4. www. epcglobalinc. org/ standards/Object_Naming_ Service_ONS_Standard_Version_1.0.pdf.

EPCglobal (2005) Guidelines on EPC for Consumer Products, Lawrenceville, NJ. www.epcglobalinc.org/public_policy/public_policy_guidelines.html.

FDA – U.S. Food and Drug Administration (2004) FDA announces new initiative to protect the U.S. drug supply through the use of radio frequency identification technology. Press release, November15. Document ID: P04-103. www.fda.gov/bbs/topics/news/2004/NEW01133.html.

FDA – U.S. Food and Drug Administration (2007) Counterfeit Colgate toothpaste found. Firm press release for immediate release, June 14. www.fda.gov/oc/po/firmrecalls/colgate06_07.htm

Feinberg R and Rousslang DJ (1990) The economic effects of intellectual property right infringements. The Journal of Business, 63(1): 79–90.

Feldhofer M, Dominikus S, and Wolkerstorfer J. (2004) Strong authentication for RFID systems using the AES algorithm. In Workshop on Cryptographic Hardware and Embedded Systems – CHES'04, Lecture Notes in Computer Science, 3156: 357–370. Berlin, Germany: Springer

Feldhofer M, Wolkerstorfer J. and Rijmen V. (2005) AES implementation on a grain of sand. In Proceedings, Information Security, 152(1): 13–20.

Finkenzeller K (1999) The RFID handbook. West Sussex, United Kingdom: John Wiley & Sons.

Finkenzeller K (2006) The RFID handbook. Munich, Germany: Carl Hanser.

Fleisch E (2001) Das Netzwerkunternehmen: Strategien und Prozesse zur Steigerung der Wettbewerbsfähigkeit in der "Networked economy". Berlin, Germany: Springer.

Fleisch E, Christ O, and Thiesse F (2005) Die betriebswirtschaftliche Vision des Internets der Dinge. In Fleisch E and Mattern F (eds.), Das Internet der Dinge: Ubiquitous Computing und RFID in der Praxis: 3–37. Berlin, Germany: Springer.

Fleisch E and Mattern F (eds.) (2005) Das Internet der Dinge: Ubiquitous Computing und RFID in der Praxis. Berlin, Germany: Springer.

Frate AA (2006) UNODC experience in estimating the global drug market. Presentation at the OECD/WIPO meeting on measurement of counterfeiting and piracy, Berne, Switzerland, October 17–18. www.oecd.org/dataoecd/42/62/35649902.pdf.

Furnham A and Valgeirsson H (2007) The effect of life values and materialism on buying counterfeit products. Journal of Socio-Economics, 36(5), pp. 677–685.

Gartner Research (2006) Market trends: PCs, Asia/Pacific, 4Q05 and year in review. Document ID: G00138769. www.gartner.com.

Gentry JW, Sanjay P, and Shultz CJII, and Suraj C (2001) How now Ralph Lauren? The separation of brand and product in a counterfeit culture. International Journal of Consumer studies, 29(6): 258–265.

Gentry JW, PutrevuS, and Shultz CJII (2006) The effects of counterfeiting on consumer search. Journal of Consumer Behaviour, 5(3): 245–256.

Glass RS and Wood WA (1996) Situational determinants of software piracy: An equity theory perspective. Journal of Business Ethics, 15(11): 1189–1198.

Givens B (2005) Activists: Communicating with consumers, speaking truth to policy makers, In: Garfinkel S and Rosenberg B (eds.), RFID: 431–437. Upper Saddle River (NJ): Addison-Wesley.

Givon M, Mahajan V, and Muller E (1995) Software piracy: Estimation of lost sales and the impact on software diffusion. Journal of Marketing, 59(1): 27–37.

Globerman S (1988) Addressing international product piracy. Journal of International Business Studies, 19(3): 497–504.

Green RT and Smith T (2002) Executive insights: Countering brand counterfeiters. Journal of International Marketing, 10(4): 89–106.

Gregory R, Lichtenstein S, and Slovic P (1993) Valuing environmental resources: A constructive approach. Journal of Risk and Uncertainty, 7(2): 177–197.

Grossman GM and Shapiro C (1988a) Counterfeit-product trade. American Economic Review, 78(1): 59–75.

Grossman GM and Shapiro C (1988b) Foreign counterfeiting of status goods. Quarterly Journal of Economics, 103(1): 79–100.

Gruen TW, Corsten DS and Bharadwaj S (2002) Retail out-of-stocks: A worldwide wxamination of extent, causes and consumer responses. Grocery Manufacturers of America, Washington D.C., USA, 2002.

GSM Association (2007) Mobile NFC services. www.gsmworld.com/documents/nfc_services_0207.pdf.

Hair JF, Anderson RE, Black B, Babin B, and Rolph EA (2005) Multivariate data analysis, 6th edition. Saddle River, NJ: Prentice Hall.

Hardgrave BC, Matthew W, and Miller R (2006) RFID's impact on out of stocks: A sales velocity analysis. http://waltoncollege.uark.edu/faculty/search.asp?type=research&id=0000445.

Harvey MG (1987) Industrial product counterfeiting: Problems and proposed solutions. The Journal of Business & Industrial Marketing, 2(4): 5–14.

Harvey M (1988) A new way to combat product counterfeiting. Business Horizons, 31(4): 19–28.

Harvey MG and Ronkainen IA (1985) International counterfeiters: Marketing success without the cost and the risk. Columbia Journal of World Business, 20(3): 37–45.

Harvey PJ and Walls WD (2003) Laboratory markets in counterfeit goods: Hong Kong versus Las Vegas. Applied Economics Letters, 10(14): 883–887.

Helpman E (1993) Innovation, imitation, and intellectual property rights. Econometrica, 61(6): 1247–1280.

Hetzler W (2002) Godfathers and pirates: Counterfeiting and organized crime. European Journal of Crime, Criminal Law and Criminal Justice, 10(4): 303–320.

Hicks JR (1970) Elasticity of substitution again: Substitutes and complements. Oxford Economic Papers, 22(3): 289–296.

Higgins RS and Rubin PH (1986) Counterfeit goods. Journal of Law and Economics, 29(2): 211–230.

Hoe L, Hogg G, and Hart S (2003) Fakin' it: Counterfeiting and consumer contradictions. European Advances in Consumer Research, 6: 60–67.

Hung CL (2003) The business of product counterfeiting in China and the post-WTO membership environment. Asia Pacific Business Review, 10(1): 58–77.

Husted BW (2004) The impact of national culture on software piracy. Journal of Business Ethics, 26(3): 197–211.

ICC – Counterfeiting Intelligence Bureau (1997) Countering counterfeiting: A guide to protecting & enforcing intellectual property rights. Document ID: 574. Paris, France: International Chamber of Commerce.

ICC – International Chamber of Commerce (2005) Current and emerging intellectual property issues for business – A roadmap for business and policy makers. Document ID: 450/911 Rev. 6. Paris, France: International Chamber of Commerce.

ICC – International Chamber of Commerce (2006) Current and emerging intellectual property issues for business – A roadmap for business and policy makers. Document ID: 450/911 Rev. 7. Paris, France: International Chamber of Commerce.

ICC – International Chamber of Commerce (2007) BASCAP online database. Information Clearing House. www.iccwbo.org/bascap/id6010/index.html.

IFPI International Federation of Phonographic Industry (2006) The recording industry 2006: Piracy report. London, United Kingdom: International Federation of the Phonographic Industry.

innovision (2006) Near field communication in the real world: Turning the NFC promise into profitable everyday applications. www.innovisiongroup.com/white_papers.php.

Jacobs L, Samli AC, and Jedlik T (2001) The nightmare of international product piracy – Exploring defensive strategies. Industrial Marketing Management, 30(6): 499–509.

Jain SC (1996) Problems in international protection of intellectual property rights. Journal of International Marketing, 4(1): 9–32.

Japanese Ministry of Internal Affairs and Communication (2007) Japan in figures 2007. Tokyo, Japan: Japanese Ministry of Internal Affairs and Communication.

Javorcik BS (2004) The composition of foreign direct investment and protection of intellectual property rights: Evidence from transition economies. European Economic Review, 48(1): 39–62.

Jenner T and Artun E (2005) Determinanten des Erwerbs gefälschter Markenprodukte – Er-gebnisse einer empirischen Untersuchung. Der Markt, 44(3/4): 142–150.

Jopling K (2005) Commercial piracy measurement. Presentation at the OECD/WIPO meeting on measurement of counterfeiting and piracy, Berne, Switzerland, October 17–18. www.oecd.org/dataoecd/43/14/35650193.pdf.

Juels A (2004) Minimalist cryptography for low-cost RFID tags. In Security in Communication Networks, Lecture Notes in Computer Science, 3352: 149–164. Berlin, Germany: Springer.

Juels A (2005a) RFID privacy: A technical primer for the non-technical reader, In: Strandburg, K. (ed.), 2005. Privacy and Identity: The Promise and Perils of a Technological Age, Springer, Berlin (to appear).

Juels A (2005b) Strengthening EPC tags against cloning. In Proceedings, 4th ACM Workshop on Wireless Security, WiSe'05: 67–76.

Juels A (2006) RFID security and privacy: a research survey. IEEE Journal on Selected Areas in Communications, 24(2): 381–394.

Kaikati JG and LaGarce R (1980) Beware of international brand piracy. Harvard Business Review, 58(2): 52–58.

Karjoth G and Moskowitz P (2005) Disabling RFID tags with visible confirmation: Clipped tags are silenced. Research Report RC23710, IBM Research Division, Zurich / Yorktown Heights (NY).

KBA – Kraftfahrt-Bundesamt (2007) Jahresbericht 2006. www.kba.de/Stabsstelle/Presseservice/ Jahrespressebericht/jpb2006.pdf

Kerlinger FN and Lee HB (1999) Foundations of behavioral research. 4th edition. Belmont, CA: Wadsworth.

Ketchen DJ Jr and Shook CL (1996) The application of cluster analysis in strategic management research. Strategic Management Journal, 17(6): 441–458.

Khouja M and Smith MA (2007) Optimal pricing for information goods with piracy and saturation effect. European Journal of Operational Research, 176(1): 482–497.

Kirkpatrick D (2007) How Microsoft conquered China – Or is it the other way around? Fortune's David Kirkpatrick goes on the road to Beijing with Bill Gates, who threw his business model out the window. CNNMoney.com, July 17, 2007.

Krechevsky C (2000) The multinational approach to anti-counterfeiting. The Journal of Proprietary Rights, 12(9): 1–12.

Langheinrich M (2005) Personal privacy in ubiquitous computing – Tools and system support, PhD thesis No. 16100, ETH Zurich, Zurich. www.vs.inf.ethz.ch/publ/papers/langheinrich-phd-2005.pdf.

Lau EK-W (2006) Factors motivating people toward pirated software. Qualitative Market Research: An International Journal, 9(4): 404–419.

Leahy P (2006) Congress passes Leahy-backed measure. Press release, March 7. http://leahy.senate.gov/press/200603/030706.html.

Leibenstein H (1950) Bandwagon, snob, and Veblen effect in the theory of consumers demand. Quarterly Journal of Economics, 64(2): 183–207.

Leisen B and Nill A (2001) Combating product counterfeiting: An investigation into the likely effectiveness of a demand-oriented approach. AMA 2001 Winter Educators' Conference, American Marketing Association Conference Proceedings (12): 271–77.

Liebowitz SJ (1985) Copying and indirect appropriability: Photocopying journals. Journal of Political Economy, 93(5): 945–957.

Liebowitzs SJ (2005) Economists' topsy-turvy view of piracy. Review of Economic Research on Copyright Issues, 2(1): 5–17.

Ling SS (2005) Piracy in the automobile sector. Asia Law, October 20. www.asialaw.com/ default.asp?page=14&ISS=20713&SID=591317.

Liu K, Li J-A, Wu Y, and La KK (2005) Analysis of monitoring and limiting of commercial cheating: A newsvendor model. Journal of Operational Research Society, 56: 877–854.

Mansfield E, Schwarz M, and Wagner S (1981) Imitation costs and patents: An empirical study. The Economic Journal, 91(364): 907–918.

Maricich A (2005) Seizure statistics. Presentation at the OECD/WIPO meeting on measurement of counterfeiting and piracy, Berne, Switzerland, October 17–18. www.oecd.org/dataoecd/43/47/35649516.pdf.

Mascarenhas B and Aaker D (1989) Strategy over the business cycle. Strategic Management Journal, 10(3): 199–210.

Maskus KE (2000) Intellectual property rights in the global economy. Institute for International Economics, Washington, DC.

McDonald G and Roberts C (1994) Product piracy – The problem that will not go away. Journal of Product & Brand Management, 3(4): 55–65.

McLean P (2006) Apple warns of phony iPods. Apple Insider, April 19. www.appleinsider.com/articles/06/04/19/apple_warns_of_phony_ipods.html.

Meyer AD, Tsui AS, and Hinings CR (1993) Configurational approaches to organizational analysis. Academy of Management Journal, 36(6): 1175–1195.

Mitchell V-W and Kearney I (2002) A critique of legal measures of brand confusion. Journal of Product & Brand Management, 11(6): 357–379.

Montgomery DC (2004) Introduction to statistical quality control. 5th edition. New York, NY: John Wiley & Sons.

Montoro-Pons JD and Cuadrado-Garcia M (2006) Digital goods and the effects of copying: An empirical study of the music market. 14th International Conference on Cultural Economics, Vienna, Austria.

Moores TT and Chang JC-J (2006) Ethical decision making in software piracy: Initial development and test of a four-component model. MIS Quarterly, 30(1): 167–180.

Moores TT and Dhaliwal J (2004) A reversed context analysis of software piracy issues in Singapore. Information & Management, 41(8): 1037–1042.

Morrison DG and Schmittlein DC (1988) Generalizing the NBD model for consumer purchases: What are the implications and is it worth the effort? Journal of Business and Economic Statistics, 6(2): 145–166.

Mullagh M (2006) Monetizing NFC solutions for mobile operators. Presentation at NFC World Asia 2006, Singapore.

NFC Forum (2006) Near field communication and the NFC forum: The keys to truly interoperable communications. www.nfcforum.org/resources/white_papers/nfc_forum_marketing_white_ paper.pdf.

Nia A and Zaichkowsky JL (2000) Do counterfeits devalue the ownership of luxury brands? Journal of Product & Brand Management, 9(7): 485–497.

Nill A and Shultz CJII (1996) The scourge of global counterfeiting. Business Horizons, 39(6): 37–42.

OECD – Organisation for Economic Co-operation and Development (2006) The economic impact of counterfeiting – Counterfeiting and piracy overall assessment. Draft for limited distribution. Paris, France: OECD.

OECD – Organisation for Economic Co-operation and Development (1998) The economic impact of counterfeiting. Paris, France: OECD. www.oecd.org/dataoecd/11/11/2090589.pdf.

Olsen JE and Granzin KL (1992) Gaining retailers' assistance in fighting counterfeiting: Conceptualization and empirical test of a helping model. Journal of Retailing, 68(1): 90–109.

Olsen JE and Granzin KL (1993) Using channels constructs to explain dealers' willingness to help manufactures combat counterfeiting. Journal of Business Research, 27(2): 147–170.

Office of the U.S. Trade Representative (2001) Continued recording piracy, U.S. suspends Ukraine's special duty-free status and issues preliminary sanctions list. Press release, July 8. Washington, DC: USTR Press Releases. www.ustr.gov/index.html.

O'Boyle J (2006) COTS and counterfeit semiconductors: Cause or effect. Mil/COTS Digest, April: 21–24.

Papadopoulos T (2004) Pricing and pirate product market formation. The Journal of Product and Brand Management, 13(1): 56–63.

Parthasarathy M and Mittelstaedt RA (1995) Illegal adoption of a new product: A model of software piracy behavior. Advances in Consumer Research (22): 693–698.

Peace GA, Galletta DF, and Thong JYL (2003) Software piracy in the workplace: A model and empirical test. Journal of Management Information Systems, 20(1): 153–177.

Penz E and Stöttinger B (2005) Forget the "real" thing – Take the copy! An explanatory model for the volitional purchase of counterfeit products. Advances in Consumer Research, 32(1): 568–576.

Phau I, Prendergast G, and Leung H (2001) Profiling brand-piracy-prone consumers: An exploratory study in Hong Kong's clothing industry. Journal of Fashion Marketing and Management, 5(1): 45–56.

Porter ME (1979) The structure within industries and companies' performance. The Review of Economics and Statistics, 61(2): 214–227.

Prendergast G, Chuen LH, and Phau I (2002) Understanding consumer demand for non-deceptive pirated brands. Marketing Intelligence & Planning, 20(7): 405–416.

Punj G and Stewart DW (1983) Cluster analysis in marketing research: Review and suggestions for application. Journal of Marketing Research, 20(2): 134–148.

Reger RK and Huff AS (1993) Strategic groups: A cognitive perspective. Strategic Management Journal, 14(2): 103–124.

Roberti M (2003) The perception question. RFID Journal, May 26, 2003. www.rfidjournal.com/ article/articleview/434/1/67/.

Ronkainen IA and Guerrero-Cusumano J-L (2001) Correlates of intellectual property violation. Multinational Business Review, 9(1): 59–65.

Sarma S, Weis S, and Engels D, (2002) RFID systems, security & privacy implications. In Lecture Notes in Computer Science, 2523: 454–469, Berlin, Germany: Springer.

Santos JF and Ribeiro JC (2006) An exploratory study of the relationship between counterfeiting and culture. Tékhne, 3(5-6): 227–243.

Schlegelmilch BB and Stöttinger B (1999) Der Kauf gefälschter Markenprodukte: Die Lust auf das Verbotene. Marketing Zeitschrift für Forschung und Praxis, 22(3): 196–208.

Shigeo S (1986) Zero quality control: Source inspection and the poka-yoke system. Portland, OR: Productivity Press.

Shultz CJII and Nill A (2002) The societal conundrum of intellectual property rights: A game-theoretical approach to the equitable management and protection of IPR. European Journal of Marketing, 36(5/6): 667–688.

Shultz C and Saporito B (1996) Protecting intellectual property: Strategies and recommendations to deter counterfeiting and brand piracy in global markets. Columbia Journal of World Business, 31(1): 18–28.

Simone JT (1999) Countering counterfeiters. The China Business Review. January/February 1999: 12–19.

Smith HJ (2001) Information privacy and marketing: What the U.S. should (and shouldn't) learn from Europe. California Management Review 43(2): 8–33.

Sonmez M and Yang D (2005) Manchester United versus China: A counterfeiting and trade-mark match. Managing Leisure, 10(1): 1–18.

Spinello RA (1998) Privacy rights in the information economy. Business Ethics Quarterly 8(4): 723–742.

Statistisches Bundesamt Deutschland (2006). Koordinierte Bevölkerungsvorausberechnung. Berlin, Germany: Statistisches Bundesamt. www.destatis.de/basis/d/bevoe/bev_svg_var.php.

Swee H, Peng S, Lim E, and Tambyah S (2001) Spot the difference: Consumer responses towards counterfeits. Journal of Consumer Marketing, 18(3): 219–235.

Swire PP (1997) Markets, self-regulation, and government enforcement in the protection of personal information. In Privacy and self-regulation in the information age: 3–19, Washington DC: U.S. Department of Commerce.

Taguchi G and Clausing D (1990) Robust quality. Harvard Business Review, 68(1): 65–75.

Tan B (2002) Understanding consumer ethical decision making with respect to purchase of pirated software. Journal of Consumer Marketing, 19(2): 96–111.

TAXUD – European Taxation and Customs Union (2001) Counterfeit and piracy: Statistical results for Germany for 2000. Brussels, Belgium European Commission http://ec.europa.eu/ taxation_customs/resources/documents/alle_2000_en.pdf.

TAXUD – European Taxation and Customs Union (2005) Counterfeit and piracy: Community-wide statistics for 2004. Brussels, Belgium: European Commission.http://ec.europa.eu/ taxation_customs/resources/documents/customs/customs_controls/counterfeit_piracy/ statistics/ counterf_comm_2004_en.pdf.

TAXUD – European Taxation and Customs Union (2006) Counterfeit and piracy: Community-wide statistics for 2005. Brussels, Belgium: European Commission. http://ec.europa.eu/ taxation_customs/resources/documents/customs/customs_controls/counterfeit_piracy/ statistics/counterf_comm_2005_en.pdf.

Thiesse F (2007) RFID, privacy, and the perception of risk: A strategic framework. In Journal of Strategic Information Systems 16(2): 214–232.

Tom G, Garibaldi B, Zeng Y, and Pilcher J (1998) Consumer demand for counterfeit goods. Psychology & Marketing, 15(5): 405–421.

Trainer (2002) The fight against trademark counterfeiting. The China Business Review, November/December: 20–24.

U.S. Customs and Border Protection (2005) Seizure statistics for intellectual property rights. Los Angeles, CA: U.S. Customs and Border Protection.

U.S. Department of Commerce, Office of Travel & Tourism Industries (2005) In-flight survey. Washington, DC: U.S. Department of Commerce.

Uncles MD, Kwok S, and Huang S (2005) Modeling retail performance using consumer panel data: A Shanghai case study. In International Conference on Services Systems and Services Management – ICSSSM'05, 1: 87–101.

UNODC – United Nations Office on Drug and Crime (2006) World drug report 2005: Analysis. 1: 125–146. New York, NY: United Nations Publications.

Veblen T (1899) The theory of the leisure class. Whitefish, MT: Kessinger Publishing.

Verhallen TMM and Henry HSJ (1994) Scarcity and preference: An experiment on unavailability and product evaluation. Journal of Economic Psychology, 15(2): 315–331.

Vigneron F and Johnson LW (1999) A review and a conceptual framework of prestige-seeking consumer behavior. Academy of Marketing Science Review, 99(1), electronic version. www.amsreview.org/amsrev/theory/vigner01-99.html.

Wagner SC and Sanders GL (2001) Considerations in ethical decision making and software piracy. Journal of Business Ethics, 29(1–2): 161–167.

Wald J and Holleran J (2007) Counterfeit products and faulty supply chain: A Johnson & Johnson Brand Integrity Case Study. Risk Management Magazine 54(4): 58.

Walsh D (2005) South San Francisco customs seizes fake Tamiflu – Nation's first haul of bogus bird flu pills traced to China. San Francisco Chronicle, Monday, Dec. 19, sec. B: 1.

Wang F, Zhang H, Zang H., and Ouyang M (2005) Purchasing pirated software: An initial examination of Chinese consumers. Journal of Consumer Marketing, 22(6): 340–351.

Wee CH, Ta SJ, and Cheok KH (1995) Non-price determinants of intention to purchase counterfeit goods – An exploratory study. International Marketing Review, 12(6): 19–46.

Wiechert TJP, Thiesse F, Michahelles F, Schmitt P, and Fleisch E (2007) Connecting mobile phones to the Internet of Things: A discussion of compatibility issues between EPC and NFC. Auto-ID Lab Working Paper No. 42. www.autoidlabs.org/uploads/media/AUTOIDLABS-WP-BIZAPP-042.pdf

Wiedemann P and Hennen L (1989) Schwierigkeiten bei der Kommunikation über technische Risiken, Arbeiten zur Risiko-Kommunikation, Issue 9, Forschungszentrum Jülich, Germany.

Wilke R and Zaichkowsky JL (1999) Brand imitation and its effects on innovation, competition, and brand equity. Business Horizons, 42(6): 9–18.

Winer RS (2001) A framework for customer relationship Mmanagement. California Management Review 43(4): 89–105.

WIPO – The World Intellectual Property Organization (2004) WIPO intellectual property handbook: policy, law and use. Document ID: 489 (E). Geneva, Switzerland: WIPO Publication. www.wipo.int/about-ip/en/iprm/index.htm.

Wolkersdorfer J (2005) Is elliptic-curve cryptography suitable to secure RFID tags? Presentation at the Workshop on RFID Security – RFIDSec'06, Graz, Austria, July 12–14. http://events.iaik.tugraz.at/ RFIDSec06/Program.

Wood AD and Stankovic JA (2002) Denial of service in sensor networks. Computer, 35(10): 54–62.

Woolley DJ and Einingen MM (2006) Software piracy among accounting students: A longitudinal comparison of changes and sensitivity. Journal of Information Systems, 20(1): 49–63.

WTO – World Trade Organization (1994) Trade-related aspects of intellectual property rights. Annex 1C of the Marrakesh Agreement Establishing the World Trade Organization, Annex 1C, Section 4, Article 51. Geneva, Switzerland: WTO Publications.

WTO – World Trade Organization (2006) International trade statistics 2005. Geneva, Switzerland: WTO Publications.

Yang D, Sonmez M, and Bosworth D (2004) Intellectual property abuses: How should multinationals respond? Long Range Planning, 37(5): 459–475.

Yao JT (2005a) Counterfeiting and an optimal monitoring policy. European Journal of Law and Economics, 19(1): 95–114.

Yao JT (2005b) How a luxury monopolist might benefit from a stringent counterfeit monitoring regime. International Journal of Business and Economics, 4(3): 177–192.

Zhang X and King B (2005) Integrity improvements to an RFID privacy protection protocol for anti-counterfeiting. In Information Security, Lecture Notes in Computer Science, 3650/2005: 474–481. Berlin, Germany: Springer.

Zimmermann R, Klein-Bölting U, Sander B, Murad-Aga T, and Bauer HH (2001) Brand equity excellence, volume 1 – Brand equity review. Düsseldorf, Germany: BBDO Group Company. www.bbdo.de/de/home/studien.download.Par.0009.Link1Download.File1Title.pdf.

List of Abbreviations

ACG	Anti Counterfeiting Group
BSA	Business Software Alliance
CASPIAN	Consumers Against Supermarket Privacy Invasion and Numbering
CEBR	Centre for Economics and Business Research
DNS	Domain Name Service
EPC	Electronic Product Code
EPC IS	EPC Information Service
EPC ONS	EPC Object Naming Service
ERP	Enterprise Resource Planning
FDA	Food and Drug Administration (USA)
ICC	International Chamber of Commerce
IFPI	International Federation of the Phonographic Industry
IP	Internet Protocol
IP	Intellectual Property
IPR	Intellectual Property Rights
OECD	Organisation for Economic Co-operation and Development
PET	Privacy Enhancing Technology
RFID	Radio Frequency Identification
TAXUD	European Taxation and Customs Union
TRIPS	Trade-Related Aspects of Intellectual Property Rights
UbiComp	Ubiquitous Computing
UNODC	United Nations Office on Drugs and Crime
URI	Uniform Resource Identifier
WIPO	World Intellectual Property Organization

Index

About the Authors

Thorsten Staake is project manager and postdoctoral fellow at the Department of Management, Technology, and Economics at the Swiss Federal Institute of Technology Zurich (ETH Zurich). Before joining ETH, Thorsten worked at MIT's Auto-ID Lab and at the Institute of Technology Management, University of St. Gallen, where he was responsible for the institute's supply chain security initiative. Thorsten received his Ph.D. in business administration from St. Gallen for a dissertation on the impact and potential countermeasures with respect to counterfeit trade, and he holds a M.Sc. in electrical engineering and information technology from Darmstadt University of Technology. He has conducted and led numerous research and consulting projects with companies from various industries. In case of questions or comments, please contact Thorsten at: tstaake@ethz.ch. Further information on the book is available at www.autoidlabs.org/ac-book.

Elgar Fleisch is professor of information and technology management at the Federal Institute of Technology in Zurich (ETH Zurich) and at the University of St. Gallen (HSG), Switzerland. His current research focuses on future developments in business computing, in particular in the area of ubiquitous computing, work that addresses the architecture and applications of the Internet of Things. Elgar Fleisch co-chairs the global network of Auto-ID Labs, and the Mobile and Ubiquitous Computing Lab (M-Lab). He is a co-founder of several university spin-offs and serves as a member in numerous steering committees in academia and business.